GRETCHEN MAYO REED,
B.S., M.A. (Ed.), M.A. (Biology)
Department of Anatomy, Physiology and Biophysics,
College of Basic Medical Sciences,
University of Tennessee Center for the Health Sciences,
Memphis, Tennessee

The late
VINCENT F. SHEPPARD,
A.B., M.A., M.Ed., Ph.D.
Learning Resource Center,
LeMoyne-Owen College,
Memphis, Tennessee

SECOND EDITION

W. B. SAUNDERS COMPANY
Philadelphia, London, Toronto

REGULATION OF FLUID AND ELECTROLYTE BALANCE:

A Programed Instruction in Clinical Physiology

W. B. Saunders Company: West Washington Square
Philadelphia, PA 19105

1 St. Anne's Road
Eastbourne, East Sussex BN21 3UN, England

1 Goldthorne Avenue
Toronto, Ontario M8Z 5T9, Canada

Library of Congress Cataloging in Publication Data

Reed, Gretchen Mayo.

Regulation of fluid and electrolyte balance.

1. Water-electrolyte imbalances—Programmed instruction. 2. Body fluids—Programmed instruction. 3. Water-electrolyte balance (Physiology)—Programmed instruction. I. Sheppard, Vincent F., 1919-1976, joint author. II. Title. [DNLM: 1. Water-Electrolyte balance—Nursing texts. 2. Water-Electrolyte balance—Programmed texts. 3. PH—Nursing texts—Hydrogen-Ion concentration—Nursing texts. 4. PH—Programmed texts—Hydrogen-Ion concentration—Programmed texts. QU18 R324r]

RC630.R43   1977   612'.01522   76-20109

ISBN 0-7216-7513-1

REGULATION OF FLUID AND ELECTROLYTE BALANCE:
A Programed Instruction in Clinical Physiology   ISBN 0-7216-7513-1

© 1977 by the W. B. Saunders Company. Copyright 1971 by W. B. Saunders Company. Copyright under the International Copyright Union. All rights reserved. This book is protected by copyright. No part of it may be reproduced, stored in a retrieval system or transmitted in any form or by any means, electronic, mechanical, photocopying, recording or otherwise, without written permission from the publisher. Made in the United States of America. Press of W. B. Saunders Company. Library of Congress catalogue card number 76-20109.

Last digit is the print number   9   8   7   6   5   4   3

PREFACE

The purpose of this Programed Instruction is to provide a self-teaching instrument for the study of fluid, electrolyte, and acid-base balance. The book is written for any member of the health team who shares responsibility for maintaining or restoring physiologic balance in the patient.

Understanding the body's mechanisms for preserving its internal environment is essential to effective management when alterations occur. It is not enough, for example, to be able to identify and administer a particular fluid intravenously. Effective therapy requires that the team know the indications for administering that fluid, its effects and side effects, and the rationale for continuing or discontinuing its administration. Omission or overextension of intravenous therapy can be, in many instances, lethal.

Basic physiological mechanisms regulate total body water and electrolyte concentrations, as well as their distribution and compartmentalization. Still other mechanisms stabilize acid-base balance. Many concepts basic to these regulatory mechanisms will have been introduced in the student's previous study of physiology from the systemic approach. This program re-examines these concepts in the context of actual clinical situations and integrates the principles underlying regulation into case studies and descriptions of disease processes.

The order of presentation of topics in the five Units of the book represents a deductive movement from the theoretical to the applied. Unit One investigates the basic concepts of fluid, electrolyte, and acid-base balance. Units Two, Three, and Four treat, in turn, fluid balance, electrolyte balance, and acid-base equilibrium. The fifth Unit presents case histories involving multiple problems which occur in the various aspects of homeostasis.

On completion of this program, a conscientious student can expect to have a working knowledge of the following specific topics:

(1) The terminology related to fluid, electrolyte, and acid-base balance.

(2) The normal distribution of fluids, ions, and molecular substances in the various compartments and the mechanisms which produce this compartmentalization.

(3) The role of the various organ systems in maintenance of internal environment.

(4) Total body water, compartmental fluid shifts, edema, dehydration, and alterations and therapy pertinent to specific disease conditions.

(5) Principles related to electrolyte balance and the pathophysiology of imbalance.

(6) Analysis of data and laboratory values related to intake and output, chemistry profile, and blood gases.

(7) The role of blood buffers, the respiratory system, and the kidneys in stabilizing pH.

(8) Identification of disease processes in which alterations in homeostasis are likely to occur.

(9) Rationale of approaches to patient care in clinical situations which alter fluids, electrolytes, and acid-base balance.

The programed method employed in this book moves by logical stages in a natural progression from the familiar to the unknown. The method enables the student to participate actively in the learning process through reading, reasoning, and writing. Repetition reinforces memory. The method is geared to maximal understanding and retention in minimal time.

The principal reference source for the theoretical content of Units One through Four was Arthur C. Guyton's *Textbook of Medical Physiology*, Fifth Edition. Illustrations used throughout the book were reproduced or adapted from figures in the same textbook with the kind permission of Dr. Guyton and the Publisher. The primary reference for Unit Five was John H. Bland's *Clinical Metabolism of Body Water and Electrolytes.* Topics for clinical applications were derived from Joan Luckmann and Karen Creason Sorensen's *Medical-Surgical Nursing: A Psychophysiologic Approach.*

The authors wish to acknowledge the expertise of Janice Schneider, B.S.N., and Bonnie Williams, B.S.N., of the Methodist Hospital School of Nursing, Memphis, Tennessee. Their critical review of the first edition and their suggestions for revision are greatly appreciated.

INTRODUCTION

Before studying the basic principles which underlie the regulation of fluid and electrolyte concentrations, it will be helpful to consider why their relatively stable state is important and to speculate on how this stability came about.

Supposedly, eons ago, when the temperature of the earth cooled sufficiently to form the oceans, these waters contained very little salt. Salinity gradually increased, however, as mineral deposits washed into the waters from the primordial continents until, with the passage of time, life appeared. Early forms of living things found the seas to be an ideal environment because their vastness prevented drastic changes in acidity, temperature, and ionic constitutents.

The high potassium content of those ancient waters was compatible with protoplasmic life within the cell and probably diffused easily across the membranes of unicellular organisms. As the sodium content of the oceans increased, this element, too, was probably incorporated as a necessary ingredient of the cellular environment. As one-celled organisms developed into multicellular forms, they enclosed within each cell fluids high in potassium and they surrounded these cells with fluids high in sodium.

If ancient seas were favorable to life, as apparently they were, animal forms may have preserved within themselves their original medium as they migrated from water to land along that long evolutionary path from unicellular structure to the complexity which is man.

Indeed, the nature of blood serum and interstitial fluids suggests that to this day every cell of the body still bathes in an internal medium identical with that of its oceanic origin.

Thus the environment of the ancient seas endures. Individual cells can live, multiply, and maintain their characteristic functions only so long as the interstitial fluids which surround them provide their optimal environmental requirements. The normal functioning of the cell demands that the composition of these fluids be relatively constant. Regulatory mechanisms maintain this constancy. An investigation of these regulatory mechanisms, and their disruption by disease, makes up the content of this book.

HOW TO USE THIS BOOK

The book is arranged in two tracks to meet the varying needs of different student groups. The main track includes all the information basic to patient care for students in accelerated programs in which time is a major consideration. The second track includes, in addition to the material in the main track, further information for the student who desires more detailed coverage in greater depth. Clear directions for proceeding are given at those places throughout the book where the additional material of Track Two is inserted.

This book divides the subject matter into five Units, each of which is subdivided into a number of Items. Each Item presents a factual explanation of a single topic, followed by several questions. The questions are of two types: fill-in and multiple-choice. The correct answer to every question appears immediately below the question. Alternative correct answers to the fill-in questions appear in parentheses.

Do not look at the answers given in the book until you have written in what you think are the correct answers. Use the piece of cardboard provided to cover the answer below each question while you write in the appropriate words in the blanks of a fill-in question, or encircle your choice of the alternative answers to a multiple-choice question. Then slide the cardboard cover down the page to reveal the correct answer, and check the accuracy of your own answer.

Avoid guessing. If you are not sure of your answer, reread the explanatory part of the Item. The need for such rereading is to be expected occasionally and perhaps even frequently. Rereading the Item will reinforce learning, eliminate vagueness, and enhance your confidence.

Turn now to page one and begin the course.

CONTENTS

PREFACE v

INTRODUCTION vii

HOW TO USE THIS BOOK ix

UNIT ONE: BASIC PHYSIOLOGY OF FLUIDS AND ELECTROLYTES 1

UNIT TWO: FLUID BALANCE 41

UNIT THREE: ELECTROLYTE BALANCE 107

UNIT FOUR: REGULATION OF HYDROGEN ION CONCENTRATION 155

UNIT FIVE: CLINICAL CORRELATIONS 217

UNIT ONE • BASIC PHYSIOLOGY OF FLUIDS AND ELECTROLYTES

DISTRIBUTION OF BODY FLUIDS

ITEM 1

Total body fluids in the average adult occupy a volume of approximately 40 liters and are referred to collectively as *total body water*. Various membranes, including cell membranes, capillary membranes, serous membranes, and mucous membranes, separate total body water into compartments.

Intracellular fluid (ICF) is that part of total body water contained within cells. *Extracellular fluid* (ECF) includes all the remainder of total body water not contained within cells. The two most important extracellular fluids are *serum* and *interstitial fluid*. Serum is the liquid component of blood, including all the dissolved and suspended particles except cells. Interstitial fluid is that fluid which immediately surrounds tissue cells. In some disease conditions, extracellular fluid accumulates in *potential spaces* in the body. Potential spaces, as their name implies, do not normally exist. They can be created, however, between membranes and within loose connective tissue when fluid accumulates as a result of fluid shifts and pressure differences. For example, the abdominal cavity can be considered a potential space—one that would become an actual space if fluid accumulated excessively in conditions which impede drainage.

The composition and activity of intracellular fluids are remarkably constant, regardless of tissue type. For this reason, intracellular fluids can be considered as a unit. Similarly, extracellular fluids are so alike that they, too, can be considered collectively, regardless of their source.

Total body water is classified by similarities in composition into two main compartments: _____ fluid and _____ fluid.

QUESTION 1

---

intracellular ... extracellular

ANSWER

Fluid contained within the cells of the body is called _____ fluid.

QUESTION 2

---

intracellular

ANSWER

Fluid not within cells is called _____ fluid, and includes primarily _____ and _____ fluid.

QUESTION 3

---

extracellular ... serum ... interstitial

ANSWER

The liquid part of blood, excluding cells, is termed _____.

QUESTION 4

---

serum

ANSWER

3

4 BASIC PHYSIOLOGY OF FLUIDS AND ELECTROLYTES

QUESTION 5
Fluid that surrounds tissue cells is called _____ fluid.

ANSWER
interstitial

QUESTION 6
In some pathological conditions, extracellular fluid can accumulate in _____ spaces within the body.

ANSWER
potential

QUESTION 7
Intracellular fluid in various types of cells throughout the body can be considered as a unit because it is similar in _____ and _____ .

ANSWER
composition . . . activity

QUESTION 8
Extracellular fluid can also be considered as a unit because it too is _____ in _____ .

ANSWER
similar . . . composition

ITEM 2 DYNAMIC EQUILIBRIUM OF BODY FLUIDS

Body fluids are normally in a state of *dynamic equilibrium*; that is, they maintain their constancy of composition in spite of an intake and output of fluid and dissolved particles which may vary drastically.

Extracellular fluids are continually mixed by the propelling action of the circulatory system and the interchange of fluid between capillaries and interstitial spaces. Fluid shifts in and out of the intracellular compartment through the semipermeable cell membranes. This perpetual state of mobility of individual molecules and ions causes a constant interchange of body fluids from one compartment to another. In spite of this continual mobility, various physiological mechanisms keep the composition of each fluid compartment within a relatively narrow normal range.

Dynamic equilibrium may be summed up as follows: although the composition of the fluid in each compartment is relatively stable, the situation as a whole is one of incessant replacement and exchange.

The situation of stable composition of fluids amidst continual mobility and interchange is called _____ _____ .	QUESTION 9
dynamic equilibrium	ANSWER
The pumping action of the heart mixes _____ fluids and promotes the interchange of fluid between _____ and _____ spaces.	QUESTION 10
extracellular ... capillaries ... interstitial	ANSWER
Fluid is exchanged within cells because of the _____ of cell membranes.	QUESTION 11
semipermeability	ANSWER
Composition of body fluids remains relatively _____ even though the composition in any one compartment is in a state of constant _____ .	QUESTION 12
stable ... change (exchange, replacement, mobility)	ANSWER

ELECTROLYTE DISTRIBUTION ITEM 3

Electrolytes are those chemical substances which dissociate in solution to form electrically charged particles, or *ions*. The same ions are present in both extracellular fluids and intracellular fluids, although the concentrations of these ions in the two compartments are strikingly different.

Intracellular fluid contains large quantities of potassium, phosphate, and magnesium and small amounts of sodium, calcium, chloride, and bicarbonate. In extracellular fluid the distribution of electrolytes is exactly reversed. Extracellular fluid has large quantities of sodium, calcium, chloride, and bicarbonate and much smaller concentrations of potassium, magnesium, and phosphate.

Figure 1 lists the concentrations of various ions and molecules in extracellular fluids and in intracellular fluids. Look closely at the distribution of the first two ions, Na^+ and K^+. Notice the striking difference in the concentration of sodium and potassium in the extracellular and intracellular compartments. Note also the difference in concentration of Cl^- and HCO_3^- in the two compartments.

The concentration of ions is given in terms of *milliequivalents per liter* (mEq./L.). A milliequivalent is the standard unit of measurement for the chemical reactivity of charged

6 BASIC PHYSIOLOGY OF FLUIDS AND ELECTROLYTES

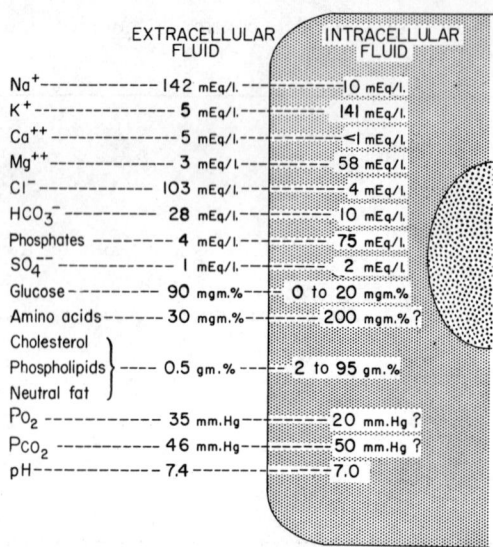

Figure 1. Chemical compositions of extracellular and intracellular fluids. (From Guyton: *Textbook of Medical Physiology*. 5th Ed. Philadelphia: W. B. Saunders Company, 1976.)

particles in body fluids. Molecules are given in *milligrams per 100 ml.* fluid (mg.%) or in *grams per 100 ml.* fluid (gm.%). Concentration of blood gases is in *millimeters of mercury* (mm. Hg). The significance of these units will be discussed in later Items.

QUESTION 13 Substances which form electrically charged particles in solution are termed _____ ; the charged particles themselves are called _____ .

ANSWER electrolytes ... ions

QUESTION 14 Extracellular and intracellular fluids differ in *kinds of/concentration of* (circle one) ions.

ANSWER concentration of

QUESTION 15 Measurement of chemical reactivity of ions in body fluids is given in

A. milligrams per 100 ml. fluid

B. milliequivalents per liter

C. millimeters of mercury

ANSWER The correct answer is B, milliequivalents per liter (mEq./L.)

Intracellular fluids contain relatively more

A. potassium, phosphate, magnesium

B. sodium, chloride, bicarbonate

QUESTION 16

The correct answer is A.

ANSWER

Extracellular fluids are high in the following electrolytes: _____, _____, and _____.

QUESTION 17

sodium ... chloride ... bicarbonate (in any order)

ANSWER

Refer to Figure 1, Item 3, and fill in the normal values in milliequivalents per liter (mEq./L.) for sodium and potassium in the following chart.

QUESTION 18

Ionic Composition of Body Fluids

Ion	mEq./L. in Extracellular Fluids	mEq./L. in Intracellular Fluids
Sodium		
Potassium		

ELECTRICAL NEUTRALITY WITHIN FLUID COMPARTMENTS

ITEM 4

In spite of markedly different distributions of the various substances among fluid compartments, the total chemical reactivity within a compartment is neutral. Within each compartment the total number of milliequivalents of *cations*, or ions with positive charges, is balanced by an equal number of milliequivalents of *anions*, or ions with negative charges. This balance of positive and negative charges is called *electrical neutrality*.

Cations are particles in solution which have _____ charges.

QUESTION 19

positive

ANSWER

8 BASIC PHYSIOLOGY OF FLUIDS AND ELECTROLYTES

QUESTION 20 Particles which are negatively charged in solution are termed _____.

ANSWER anions

QUESTION 21 In both compartments, cations balance anions in terms of chemical _____.

ANSWER reactivity

QUESTION 22 Electrical neutrality within each compartment means that the total electrical charges of _____ balance the total electrical charges of _____.

ANSWER cations ... anions

ITEM 5 COMPOSITION OF FLUIDS IN mEq./L.

Figure 2 illustrates the balance between cations and anions in serum (plasma), interstitial fluid, and cell fluid. Notice that when given in milliequivalents per liter, positive ions are balanced by negative ions. Total cations, in milliequivalents, equal total anions, again in milliequivalents.

Although the total chemical reactivity, or combining power, of cations and anions is equal within a given compartment, the total number of ions is not necessarily equal. The total number of cations compared to the total number of anions in the same compartment does not constitute electrical neutrality. Some ions, positive or negative, have two or three times the combining power of others. Neutrality is the result of the total balance of chemical reactivity.

QUESTION 23 Both intracellular and extracellular compartments are approximately equal in

A. total number of positive and negative ions

B. total chemical reactivity due to cations and anions

ANSWER A. This answer is incorrect. Each compartment does not contain an equal number of positive and negative ions. Rather, there is a balance of positive charges against negative charges in the same compartment, as far as chemical reactivity is concerned. Reread the question, then read Answer 23-B.

B. This answer is correct. In each compartment, the total reactivity of cations is balanced by an equivalent reactivity of anions in the same compartment. Thus, as far as chemical equivalence is concerned, there is electrical balance in each compartment.

COMPOSITION OF FLUIDS IN mEq./L. 9

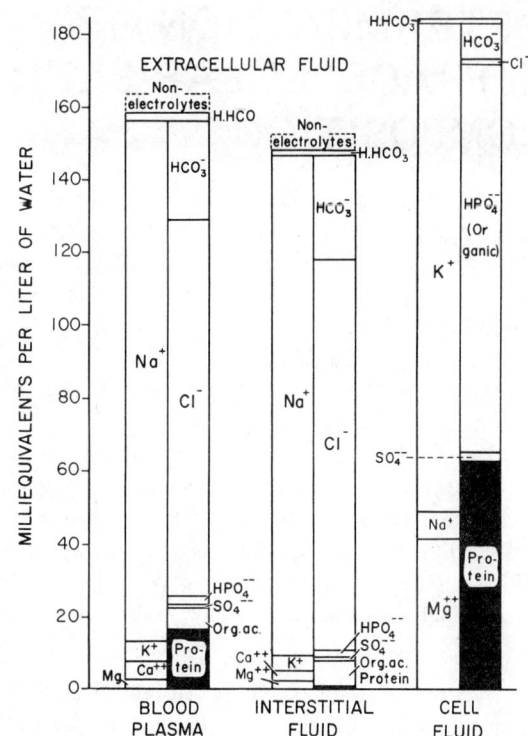

Figure 2. The compositions of plasma, interstitial fluid, and intracellular fluid. (Modified from Gamble: Chemical Anatomy, Physiology, and Pathology of Extracellular Fluid: A Lecture Syllabus. Harvard University Press, 1954. *In* Guyton: *Textbook of Medical Physiology.* 5th Ed. Philadelphia: W. B. Saunders Company, 1976.)

Refer to the bar graph in Figure 2. Notice that the three columns represent the electrolytes and non-electrolytes in blood (serum), interstitial fluid (that fluid which immediately surrounds cells), and cell fluid. Notice, too, that within a compartment (for example, plasma) the bar which represents cations is balanced by the bar which represents anions when the units of measure are given in

QUESTION 24

A. number of ions

B. milliequivalents per liter

A. This answer is incorrect. In any electrically neutral solution, the *total number of cations* may or may not be the same as the *total number of anions*, depending on the valence. Read Answer 24–B.

ANSWER

B. This answer is correct. When given in milliequivalents per liter, the bar representing cations equals the bar representing anions in an electrically neutral solution. This unit of measure, milliequivalents per liter, must recognize the difference in valence, or combining power, or chemical reactivity, between ions.

TRACK TWO. Items 6 through 13 deal chiefly with details of units of measure. Students in accelerated programs doing Track One may omit these Items and go on directly to Item 14.

ITEM 6 — DETERMINATION OF INTRACELLULAR FLUID COMPOSITION

Because it is so readily available, serum is the extracellular fluid utilized to determine the composition of body fluids. Knowing serum sodium levels, for example, combined with other information, provides a reliable estimate of intracellular sodium as well. Intracellular chloride, likewise, is routinely determined from serum chloride levels. Again, extracellular bicarbonate level, calculated from carbon dioxide content of serum, is a reliable index of intracellular bicarbonate.

The extracellular level of potassium is not so reliable an indication of intracellular potassium. Under normal conditions, it is true, serum potassium together with other information about the patient may give some index of potassium within cells. But when abnormalities are present, extracellular potassium can be a deceptive measure of intracellular potassium because of various processes which specifically regulate potassium shifts. Clinical situations involving tissue trauma, malnutrition, or insulin injections can affect cell membrane transport of potassium so that serum levels suggest estimates of intracellular potassium which may, in fact, be false highs or false lows.

QUESTION 25: Determination of serum electrolyte levels is sometimes an unreliable indication of intracellular _____ levels because of various mechanisms which regulate fluid and electrolyte _____ .

ANSWER: potassium ... shifts

QUESTION 26: Interpretation of serum potassium levels is especially hazardous in such clinical situations as tissue _____ , _____ injection, or _____ .

ANSWER: trauma ... insulin ... malnutrition

ITEM 7 — SOME UNITS OF MEASURE

At this point in the program it will be helpful to review various units of measure and their symbols. The symbol *pH* signifies the negative logarithm of the hydrogen ion concentration. A high hydrogen ion concentration, as in *acidosis*, is indicated by a low pH. Conversely, a relatively low hydrogen ion concentration, as in *alkalosis*, is indicated by a high pH. The significance of pH will be dealt with in detail in Unit Four.

The symbol *mm. Hg* stands for millimeters of mercury, the standard measure of pressure (e.g., blood pressure) necessary to raise a column of mercury to a height measured in millimeters.

The symbol *gm.%* means grams per 100 milliliters of fluid. The symbol *mg.%* means milligrams per 100 milliliters of fluid. Since body fluids are relatively dilute, concentrations of most dissolved particles in extracellular and intracellular fluids are in milligrams per 100 milliliters of fluid, or mg.%.

A low hydrogen ion concentration is indicated by a   QUESTION 27

A. low pH

B. high pH

A. This answer is incorrect. Read Answer 27–B.   ANSWER

B. This answer is correct. For most purposes, hydrogen ion concentration is expressed in terms of pH. Since pH refers to the *negative* logarithm of the hydrogen ion concentration in mEq./L., a low pH means a high concentration, and a high pH means a low concentration.

The relationship between a milligram and a gram is that a milligram is (A) *1/1000 of a gram* or (B) *1000 grams*.   QUESTION 28

A is correct — 1/1000 of a gram.   ANSWER

MEASUREMENT OF ELECTROLYTE CONCENTRATIONS   ITEM 8

Instead of measuring electrolyte concentrations in units of weight, the combining power of ions is quantified in terms of *chemical reactivity*, or *equivalents*. An equivalent is the amount of a substance that will combine with one atomic weight of hydrogen. A *milliequivalent* is 1/1000 of an equivalent. The chemical reactivity of electrolytes in body fluids is expressed in *milliequivalents per liter*, abbreviated mEq./L. Using milliequivalents permits the use of small whole numbers to indicate how many reactive units of any ion are present in a solution.

12 BASIC PHYSIOLOGY OF FLUIDS AND ELECTROLYTES

QUESTION 29

Serum electrolyte concentrations are given in terms of

A. mg.%

B. mEq./L.

C. mm. Hg

ANSWER

A. This answer is incorrect. Milligrams per 100 milliliters are units for measuring glucose, amino acids, and many other non-ionized substances in serum, but not for measuring electrolyte concentrations.

B. This answer is correct. Milliequivalents per liter is a measure of the chemical reactivity, or combining power, of ions.

C. This answer is incorrect. Millimeters of mercury are used as the unit of measure for pressure relationships.

QUESTION 30

The primary advantage of quantifying values in milliequivalents is that the method indicates, in small whole numbers,

A. the combining power of ions present

B. the milligram weights of ions present

ANSWER

A. This answer is correct. Milliequivalents tell how many combining, or reactive, units of an element are present in body fluids. The reactivity of one ion with another is what is important to know, from the standpoint of biochemical reactions.

B. This answer is incorrect. It is true that most of these values are small whole numbers, but this fact is not the main reason for quantifying values in milliequivalents. Read Answer 30–A.

ITEM 9 COMPUTING MOLECULAR WEIGHT

Milliequivalents are the units of measure for chemical reactivity, or combining power, of various ions in body fluids. A review of the chemical computation of molecular weights is helpful in understanding how milliequivalents are related to weight.

The *atomic weight* of oxygen, you will recall from general chemistry, is arbitrarily set at 16. The atomic weights of other atoms are numbers expressing the weight of any given atom relative to the weight of oxygen. The atomic weights of the various elements are found in a periodic table of the elements inlcuded in any standard chemistry textbook. The atomic weight of hydrogen, for instance, is 1.008; carbon, 12.011; sodium, 22.991; calcium, 40.008; potassium, 39.100; chlorine, 35.453.

Molecular weight is the sum of the atomic weights of all atoms present in a molecule of a given compound. The molecular weight of water is the sum of the atomic weights of two atoms of hydrgen and one atom of oxygen, or (2 X 1.008) + 16, a total of 18.016, or, expressed as a whole number, 18.

The atomic weight of sodium is 22.9; the atomic weight of chlorine is 35.4. The molecular weight of sodium chloride is _____.	QUESTION 31

58.3   ANSWER

In rough numbers, the atomic weight of calcium is 40; of chlorine, 35. The formula for calcium chloride is $CaCl_2$. What is the molecular weight of calcium chloride? Remember, molecular weight is the sum of atomic weights as they appear in a formula. A. 110 B. 75	QUESTION 32

A. This answer is correct. The atomic weight of calcium is 40; of chlorine, 35. Substituting the atomic weights of calcium and chlorine in the formula for calcium chloride, $CaCl_2$, the molecular weight is $40 + (2 \times 35)$, or $40 + 70$, which equals 110.   ANSWER

B. This answer is incorrect. This answer is the result of mistakenly adding simply the atomic weights of calcium and chlorine: $40 + 35 = 75$. Calcium chloride contains *two* atoms of chlorine, so the computation should have been $40 + (2 \times 35) = 110$.

GRAM MOLECULAR WEIGHT (MOL)   ITEM 10

When molecular weight is expressed in grams, it is called *gram molecular weight* and is abbreviated *mol*. A mol of a substance, then, is the molecular weight of that substance in grams. For example, 1 mol of $CaCl_2$ is 110 grams $CaCl_2$. A *millimol* (mM.) is 1/1000 of a mol. A millimol of $CaCl_2$ is 110 milligrams.

We can readily imagine one mol of $CaCl_2$ as 110 grams of that substance by weight. Dissolve that quantity in water to make a liter of solution and there is still 1 mol, but it is in a solution having a concentration that is expressed as 1 mol per liter.

One mol of sodium chloride (NaCl) equals the sum of the atomic weights of sodium and chlorine (23 + 35), or 58 grams. Put this one mol of sodium chloride into one liter of water and the solution contains 1 mol per liter (1 mol/L.) of NaCl.

A mol is the _____ of a substance expressed in grams.	QUESTION 33

molecular weight   ANSWER

The measure 1 mol/L. equals one gram _____ per liter of solution.	QUESTION 34

molecular weight   ANSWER

ITEM 11   RELATING MOLS TO EQUIVALENTS

The term *equivalent*, you will recall, refers to that weight of a substance which will combine with one atomic weight of hydrogen. *Gram equivalent weight* is an equivalent expressed in grams. To ascertain the equivalent weight of an ion, divide the atomic weight of that ion, in grams, by the valence, or charge. Thus, for sodium chloride, one equivalent of sodium equals 23 grams of sodium divided by the valence, which in this case is one. One equivalent of chloride equals 35 grams of chloride divided by the valence, which again happens to be one. Putting these data together, we conclude that 58 grams of NaCl (23 gm. Na + 35 gm. Cl) comes to one mol. This 1 mol contains, in terms of reactivity, one equivalent of sodium and one equivalent of chloride.

QUESTION 35   If one mole of NaCl (23 + 35 = 58) contains 1 Eq. Na^+ and 1 Eq. Cl^-, how many equivalents of calcium are in one mole of $CaCl_2$ (110 gm.)?

A. 2

B. 1

C. 3

BEFORE CHOOSING YOUR ANSWER, RECALL THE DEFINITION OF AN EQUIVALENT: *One equivalent weight of an ion equals the atomic weight of that ion in grams divided by the valence, or charge.*

ANSWER   A. This answer is correct. The amount of calcium in one mol of $CaCl_2$ is 40/110 of the total molecular weight of this compound. Since an equivalent is the atomic weight divided by the valence, or 40/2 in this case, then 40 grams of calcium equals 2 equivalents.

Suppose we take 1 mol of calcium carbonate ($CaCO_3$), molecular weight 100. The valence of the carbonate radical is 2. It combines equally with the calcium, equivalent to equivalent. Thus 2 Eq. of calcium (40 gm.) combines with 2 Eq. of carbonate (60 gm.)

B. This answer is incorrect. This example requires a look at valence, or charge. The valence of calcium is 2, whereas the valence of sodium is 1. Recall the definition of an equivalent: *the combining power of an ion contained in an amount equal to the gram atomic weight of that ion, divided by the valence.* In the case of sodium, 1 Eq. equals the atomic weight of sodium divided by the valence, or 23/1, or 23 grams. Twenty-three grams of sodium, then, is 1 Eq. of sodium. In 1 mol of sodium chloride, 23/58 of this weight is sodium—1 Eq. of sodium.

The same line of reasoning applies to calcium. One equivalent of calcium equals the atomic weight (40) divided by the valence (2). Therefore, 20 grams of calcium is one equivalent of calcium. But one mol of calcium chloride (110 grams) contains 40 grams of calcium. It also contains 2 equivalents of chloride.

C. This answer is incorrect. $CaCl_2$ dissociates into 3 ions, it is true, but this fact is irrelevant to the question.

COMPUTING REACTIVE UNITS ITEM 12

Chemical reactivity of a substance depends on the ionic charge and on the concentration. Determination of chemical reactivity is fairly simple in a substance as uncomplicated as sodium chloride, but it gets a bit involved in a solution such as serum, which contains many different kinds of electrolytes.

Consider this illustration: If you add 1 mM. sodium bicarbonate (84 mg. $NaHCO_3$) to a liter of solution containing 1 mM. sodium chloride, how many reactive units of sodium do you have? Of chloride? Of bicarbonate? One millimol (mM.) of sodium chloride contains one milliequivalent (mEq.) of sodium and one milliequivalent (mEq.) of chloride. One mM. of $NaHCO_3$ contains 1 mEq. of sodium and 1 mEq. of bicarbonate. The mixed solution of NaCl and $NaHCO_3$ contains 2 mEq. of sodium, 1 mEq. of chloride, and 1 mEq. of bicarbonate. Notice how the total mEq. of cations (2 of sodium) equals the total mEq. anions (one each of chloride and bicarbonate)—maintaining electrical neutrality in the fluid. This use of milliequivalents permits the comparison, in terms of reactivity, of the concentration of each ion.

Milligrams differ from milliequivalents and from millimols. Milligrams are units of _____ . **QUESTION 36**

---

weight **ANSWER**

Millimols are units of _____ weight and vary with the atomic weight of each compound. **QUESTION 37**

---

molecular **ANSWER**

Milliequivalents are units of _____ of individual ions, and depend on _____ and _____ . **QUESTION 38**

---

chemical reactivity . . . charge . . . concentration **ANSWER**

When 1 mM. NaCl and 1 mM. KCl are added to 1 liter of water, the mixed solution contains _____ mEq. of sodium, _____ mEq. of potassium, and _____ mEq. of chloride. **QUESTION 39**

---

one . . . one . . . two **ANSWER**

16 BASIC PHYSIOLOGY OF FLUIDS AND ELECTROLYTES

QUESTION 40 Any one fluid compartment contains a mixture of ions. The total number of mEq. of all positive ions, or _____, is equal to the total number of mEq. of all negative ions, or _____.

ANSWER cations ... anions

QUESTION 41 Refer to Figure 2. The cationic component of serum (plasma) is largely due to _____, whereas the anionic component is due mostly to _____.

ANSWER sodium (Na^+) ... chloride (Cl^-)

QUESTION 42 In cell fluid, the cationic component is largely due to _____, whereas the anionic component is due to _____.

ANSWER potassium (K^+) ... phosphate ($PO_4^=$)

QUESTION 43 Total cations in each compartment, in milliequivalents, is _____ to total anions, in milliequivalents, in the same compartment.

ANSWER equal

ITEM 13 QUANTIFYING OSMOLARITY

Total solute concentration of body fluids is represented in units of *osmolarity*. The *osmol* is the unit of measure for quantifying osmolarity. In quantifying the osmotic pressure of body fluids, it is convenient to use a smaller unit, the *milliosmol* (abbreviated mOsm.), which equals 1/1000 of an osmol.

The number of milliosmols is determined by the total number of particles in the solution, whether molecules or ions. Each particle has a unit value of 1, regardless of its size, in the case of molecules, or of its charge, in the case of ions. If a substance ionizes, each ion contributes the same amount as one non-ionizable molecule; namely, a unit value of 1. Thus, for example, one mol of glucose, a non-ionizable molecule, has an osmolar value of 1; one mol of potassium chloride (KCl), because it forms two ions, has an osmolar value of 2; one mol of calcium chloride ($CaCl_2$), because it forms three ions, has an osmolar value of 3.

The normal solute concentration of body fluids is approximately 300 milliosmols per liter (300 mOsm./L.).

Osmolarity means the _____ of solute in a solution.

QUESTION 44

concentration

ANSWER

The unit of measure for quantifying osmolarity of body fluids is the _____, which equals 1/1000 of an osmol.

QUESTION 45

milliosmol

ANSWER

In terms of osmolarity

A. both number and charge of particles are relevant

B. number of particles alone is relevant

C. charge of particles alone is relevant

QUESTION 46

The correct answer is B. Osmotic pressure of body fluids, measured in milliosmols, depends entirely upon the number of particles in solution and has nothing whatever to do with valence or charge. In terms of equivalency, on the other hand, chemical reactivity is determined by both the total number of ions and the total number of charges on each ion.

ANSWER

Solute concentration of body fluids is _____ mOsm./L.

QUESTION 47

300

ANSWER

Students taking the accelerated Track One proceed with Item 14.

HOMEOSTASIS ITEM 14

The regulation of fluid and electrolyte balance consists of (a) preservation of optimum total body concentration of water and dissolved substances and (b) distribution of chemically active particles between intracellular and extracellular fluids. Normal functioning of the cell demands that the composition of interstitial fluid be relatively constant. The term *homeostasis* is used to describe the constancy of composition of interstitial fluid necessary for cell survival. Many physiologic processes and most of the organ systems are involved in maintenance of homeostasis, or the regulation of fluid, electrolyte, and acid-base balance.

QUESTION 48
Regulation of fluid and electrolyte balance is concerned with total body _____ of substances and with _____ of these substances between compartments.

ANSWER
concentration ... distribution

QUESTION 49
The constancy of composition of interstitial fluid necessary for cell survival and optimum function is called _____.

ANSWER
homeostasis

QUESTION 50
Maintaining homeostasis, or regulation of fluid and electrolyte balance, is a function of most _____ _____.

ANSWER
organ systems

ITEM 15   THE ROLE OF THE KIDNEYS

The organs most obviously essential to regulation of fluid and electrolyte balance are the kidneys. The kidneys control the total optimal quantity of fluid and electrolytes within the body as a whole. By means of numerous hormonal mechanisms, the kidneys excrete varying amounts of water and reabsorb or release varying quantities of sodium, potassium, chloride, bicarbonate, and hydrogen ions, all in terms of regulating optimal intracellular and extracellular concentrations.

QUESTION 51
The principal determinant for kidney excretion or retention of various substances is the maintenance of optimum _____ and _____ levels of these substances.

ANSWER
intracellular ... extracellular

THE ROLE OF THE CARDIOVASCULAR SYSTEM

ITEM 16

The cardiovascular system also helps maintain balance at the cellular level. It does so by means of *capillary exchange mechanisms*. Ions and small molecular weight molecules, along with fluid, diffuse continuously from capillaries into interstitial spaces, and back from interstitial fluids into the distal end of capillaries—providing a constantly interchanged environment for cells.

The cardiovascular system participates in fluid and electrolyte distribution by means of _____ mechanisms.

QUESTION 52

capillary exchange

ANSWER

Substances which diffuse back and forth between capillaries and interstitial spaces include _____, _____, and small molecular weight _____.

QUESTION 53

ions ... fluid ... molecules

ANSWER

THE ROLE OF THE GASTROINTESTINAL SYSTEM

ITEM 17

Normal gastrointestinal function affects fluid balance because of involvement in total intake and output. Gastrointestinal disturbances such as vomiting or diarrhea can affect total body water and electrolyte concentrations adversely. Vomiting hydrochloric acid secreted by the gastric mucosa removes total acid from the body. If this condition is prolonged, metabolic alkalosis may result. This alkalosis in turn affects other mechanisms of fluid and electrolyte balance. Diarrhea results in excessive loss of both total body water and electrolytes, as absorption is impaired. Since fluids secreted into the duodenum are alkaline, excessive loss of these fluids in diarrhea can result in metabolic acidosis. Acidosis, like alkalosis, affects other control mechanisms.

The gastrointestinal system affects fluid and electrolyte balance as far as total _____ and _____ are concerned.

QUESTION 54

intake ... output

ANSWER

BASIC PHYSIOLOGY OF FLUIDS AND ELECTROLYTES

QUESTION 55 Vomiting of _____ secreted by the gastric mucosa depletes the body's total _____.

ANSWER hydrochloric acid ... acid

QUESTION 56 Diarrhea results in excessive loss of both total body _____ and _____.

ANSWER water ... electrolytes

QUESTION 57 Because it results in loss of hydrochloric acid, prolonged vomiting can bring on metabolic _____.

ANSWER alkalosis

QUESTION 58 Because of loss of alkaline duodenal secretions, prolonged diarrhea can result in metabolic _____.

ANSWER acidosis

ITEM 18   THE ROLE OF THE ENDOCRINE SYSTEM

The endocrine system is conspicuously involved in fluid and electrolyte balance. Hormones secreted by the endocrine system play an important role in the conservation and distribution of various ions and total body water. For example, aldosterone, secreted by the adrenal cortex, causes renal reabsorption of sodium. Aldosterone secretion is increased when the body needs to conserve sodium; it is decreased when the body needs to excrete sodium.

QUESTION 59 The endocrine system controls conservation of varous ions and total body water by the action of _____.

ANSWER hormones

Aldosterone is secreted by the _____.	QUESTION 60
adrenal cortex	ANSWER
Varying levels of aldosterone control the renal reabsorption of _____.	QUESTION 61
sodium	ANSWER
Increased secretion of aldosterone *increases/decreases* (circle one) renal reabsorption of sodium, so that total body sodium levels are *conserved/depleted* (circle one).	QUESTION 62
increases ... conserved	ANSWER

INTERRELATIONSHIP OF SYSTEMS

ITEM 19

Since many organ systems are involved in *homeostasis*, the constancy of the internal environment, it follows that malfunction of any one of these systems can cause disturbance of balance. Therapy is directed, then, at reestablishing this balance until the diseased system can return to normal function. Return to normalcy is achieved by intravenous therapy, peritoneal dialysis, use of an artificial kidney, regulation of intake, and other expedients.

The preceding Items dealt with the roles of the kidney, cardiovascular system, gastrointestinal system, and the endocrine system, respectively, in maintenance of fluid and electrolyte balance. But these systems do not affect homeostasis independently of one another. Their collective effect is what is meant by *interrelationship of systems* in maintaining or disturbing balance.

The following phenomenon exemplifies the interrelationships of systems. Aldosterone, secreted by an endocrine gland, exerts a direct effect on the tubules of the kidney to bring about reabsorption of sodium. Reabsorption of sodium, in turn, results in retention of water and the consequent increase in blood volume. Change in blood volume then affects the functioning of the cardiovascular system.

22 BASIC PHYSIOLOGY OF FLUIDS AND ELECTROLYTES

QUESTION 63

Homeostasis refers to

A. the relatively stable composition of fluid in the intracellular and extracellular compartments

B. a constant concentration of water and solutes in the extracellular compartment

ANSWER

A. This answer is incorrect. The relative stability of fluid composition throughout a perpetual interchange of fluid from one compartment to the other is dynamic equilibrium, not homeostasis.

B. This answer is correct. Homeostasis refers to constancy of the internal environment of cells; that is, constant concentration in the extracellular fluid.

QUESTION 64

Since the functioning of one organ system affects the functioning of other organ systems in terms of homeostasis, the various systems are said to be _____.

ANSWER

interrelated

ITEM 20 DISEASE CONDITIONS DISTURBING BALANCE

In many hospital situations, clinical disorders bring on water and electrolyte abnormalities of major concern. Sometimes fluid and electrolyte disturbances result from trauma. Burns which cover a major portion of the body surface cause a loss of extracellular fluid and protein throughout the denuded areas of the skin. Fluid shifts from the intracellular compartment to compensate for the extracellular loss. A shift of potassium accompanies this shift of intracellular fluid. The resultant *hyperkalemia* (high serum potassium) adversely affects the normal rhythm of the heart.

Hemorrhagic shock diminishes blood volume, venous return, cardiac output, and blood pressure. All these hypovolemic implications jeopardize the normal capillary exchange of fluid between the vascular system and the interstitial spaces. Among the adjustments which immediately follow hemorrhage is a shift of fluid from the interstitial spaces into the vascular system, as capillary dynamics attempt to compensate.

Various conditions result in a reduction of fluid intake drastic enough to impair water balance. Such conditions include coma, senile debilitation, and damage to the thirst center of the brain.

QUESTION 65

Many disease conditions require intake regulation by intravenous therapy and other hospital procedures in order to maintain fluid and electrolyte balance. Among these disease conditions are the following: (1) _____, (2) _____, and (3) reduction of _____ intake.

ANSWER

trauma (or burns) ... hemorrhage ... fluid

Burns cause a loss of _____ fluid and a compensating shift of _____ fluid.	QUESTION 66
extracellular . . . intracellular	ANSWER
Hypovolemia resulting from hemorrhagic shock brings about shifts of fluid from _____ space into the _____ system.	QUESTION 67
interstitial . . . vascular	ANSWER

TRANSPORT OF SUBSTANCES THROUGH MEMBRANES ITEM 21

Concentration of body fluids and distribution of body fluids between plasma, interstitial spaces, and the cell involve transport across various membranes. These membranes also provide partitions across which electrolytes are distributed in varying concentrations. A review of the structure of a typical membrane will be helpful in understanding the various processes which govern transport across the membrane.

Figure 3 illustrates the assumed arrangement of a typical membrane. Note that a single layer of lipid substance is bounded on both sides by protein molecules. A layer of mucopolysaccharide adheres to the outer protein layer. The lipid, protein, and polysaccharide molecules which make up these cell membranes are arranged in such a way that the membrane is *permeable* to some substances and *impermeable* to others. *Pores* in the membrane allow the passage of water-soluble substances of low molecular weight. Lipid-soluble substances such as oxygen and carbon dioxide, by virtue of their solubility, simply enter the lipid substance of the membrane at one side and leave from the other side.

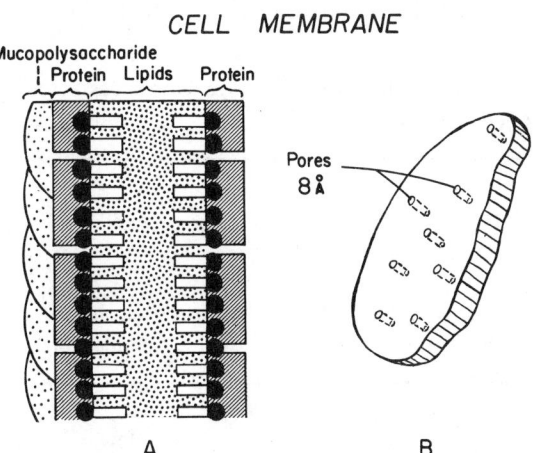

Figure 3. (A) Postulated molecular organization of the cell membrane. (B) Pores in the cell membrane. (From Guyton: *Textbook of Medical Physiology.* 5th Ed. Philadelphia: W. B. Saunders Company, 1976.)

24 BASIC PHYSIOLOGY OF FLUIDS AND ELECTROLYTES

QUESTION 68 Cell membranes are made up of _____, _____ and _____.

ANSWER lipid ... protein ... polysaccharides

QUESTION 69 Small water-soluble particles diffuse through _____ in the membrane.

ANSWER pores

QUESTION 70 Lipid-soluble particles diffuse through the _____ substance of the membrane.

ANSWER lipid

ITEM 22 PROCESSES OF TRANSPORT

Whether or not any substance, water-soluble or lipid-soluble, actually goes across a membrane depends on the combined action of several processes. Diffusion and osmosis move substances from high to low concentration areas and tend to equalize body fluids as they are distributed across a membrane. *Diffusion* refers to movement of a gas, a liquid, or a solute particle. *Osmosis* applies specifically to the movement of water. *Active transport* moves substances in the opposite direction, from an area of low concentration to one of high concentration. Active transport moves substances up a concentration gradient and tends to create composition differences.

QUESTION 71 Diffusion and osmosis move substances across a membrane from an area of _____ concentration to an area of _____ concentration.

ANSWER high ... low

QUESTION 72 Active transport moves substances across a membrane from an area of _____ concentration to an area of _____ concentration.

ANSWER low ... high

TRANSPORT OF SUBSTANCES BY DIFFUSION
ITEM 23

Diffusion is the movement of a substance from an area of high concentration to an area of low concentration, with energy derived from the normal kinetic energy of matter. Diffusion explains the translocation of water soluble substances whose molecular weight is low enough for these substances to move through the pores of the membrane. Diffusion also accounts for translocation of lipid-soluble substances such as oxygen and carbon dioxide through the lipid portion of the membrane itself. Diffusion effects the movement of substances from plasma, through capillary membranes, into interstitial fluids, and across cell membranes. The direction of net transfer is always toward equalizing concentrations.

Small-sized water-soluble particles diffuse through membrane _____. — QUESTION 73

pores — ANSWER

Oxygen and carbon dioxide diffuse through the _____ portion of the cell membrane. — QUESTION 74

lipid — ANSWER

The process of diffusion _____ concentration differences. — QUESTION 75

equalizes — ANSWER

CONCENTRATION GRADIENT AND ELECTRICAL GRADIENT
ITEM 24

Net diffusion is influenced by the *concentration gradient*, the *electrical gradient*, and the *pressure gradient*, as illustrated in Figure 4. A concentration gradient is the difference in concentration on the two sides of the membrane. Part A of Figure 4 shows the movement of particles from a densely populated area to a less dense area.

An electrical gradient is that force created by a difference in the number of positive and negative charges across the membrane. Unlike ions attract and move toward each other. Ions which have similar charges are repelled. Part B of Figure 4 shows, on the left,

A. CONCENTRATION DIFFERENCE

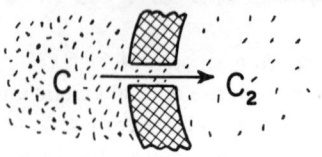

B. ELECTRICAL POTENTIAL DIFFERENCE

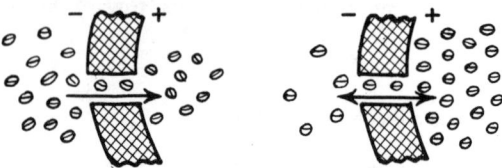

Figure 4. Effect of (A) concentration difference, (B) electrical difference, and (C) pressure difference on diffusion of molecules and ions through a cell membrane. (From Guyton: *Textbook of Medical Physiology.* 5th Ed. Philadelphia: W. B. Saunders Company, 1976.)

C. PRESSURE DIFFERENCE

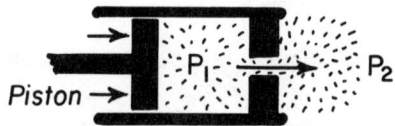

the initial situation of movement of negative ions from the extracellular compartment across a membrane into intracellular fluids which are predominantly positive. On the right of Figure 4, Part B, notice how the nondiffusible positive ions in the intracellular fluid create a condition in which diffusible negative ions distribute unequally to maintain electrical neutrality.

An additional factor is operative in an electrical gradient: membrane pores are lined with positive charges, primarily because of calcium. Positive charges within the pore tend to impede passage of positive ions, while allowing negative ions to pass with ease.

QUESTION 76: Net diffusion is determined by the _____ gradient, the _____ gradient, and the _____ gradient.

ANSWER: concentration . . . electrical . . . pressure

QUESTION 77: Differences in concentration of substances on either side of a membrane are referred to as a _____ gradient.

ANSWER: concentration

QUESTION 78: Differences in positive and negative charges across a membrane are called an _____ gradient.

ANSWER: electrical

Since the pores of a membrane are lined with positive charges, substances with a _____ charge can pass through more easily.

QUESTION 79

negative

ANSWER

PRESSURE GRADIENT ITEM 25

A pressure gradient, illustrated in Figure 4, Part C, is the difference between the pressures on the two sides of a membrane. When the pumping action of the heart, for example, imposes a hydrostatic pressure gradient on the arteriole side of a capillary, fluid moves from the capillary into interstitial fluids. On the venule side of the capillary, this pressure is reduced to such a degree that direction of flow is reversed from interstitial fluid to plasma.

Differences in pressure on either side of a membrane are termed a _____ _____.

QUESTION 80

pressure gradient

ANSWER

CAPILLARY EXCHANGE ITEM 26

Figure 5 illustrates the diffusion of fluids back and forth across the capillary membrane and through the interstitial spaces. The longer arrows at the arteriole end of

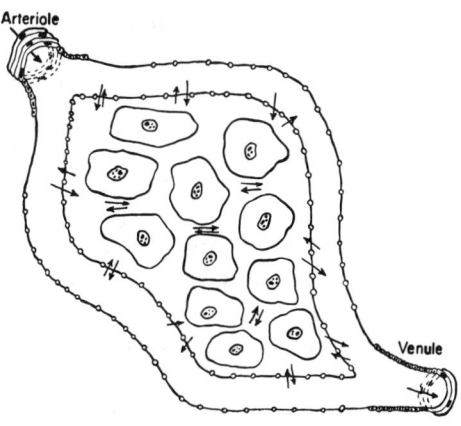

Figure 5. Diffusion of fluids through the capillary walls and through the interstitial spaces. (From Guyton: *Textbook of Medical Physiology.* 5th Ed. Philadelphia: W. B. Saunders Company, 1976.)

the capillary bed indicate a gradient favoring diffusion from blood to interstitial spaces. Notice at the venule side of the capillary bed that longer arrows indicate factors favoring diffusion in the opposite direction, from interstitial fluid into blood. Notice also in Figure 5 the close proximity of any cell to its site of nutrient supply, the capillary.

QUESTION 81

Diffusion from interstitial fluid into blood occurs at the _____ side of the capillary bed; diffusion from blood into interstitial fluid, at the _____ end of the capillary bed.

ANSWER

venule ... arteriole

ITEM 27 DIALYSIS

An important clinical application of the principle of diffusion is the use of *dialysis*, a procedure designed to remove excess electrolytes and toxic substances from body fluids. Patients with kidney diseases which prevent excretion of these excesses respond quickly to dialysis as a means of adjusting serum concentrations of dissolved substances.

Dialysis is the differential diffusion of certain particles through a semipermeable membrane. The pores of the dialysis membrane permit small particles, such as ions, to diffuse, but restrain larger sized molecules, such as serum proteins and blood cells.

In the procedure known as peritoneal dialysis the peritoneal membrane serves as the semipermeable dialysis membrane. A fluid which artificially simulates extracellular fluid is introduced into the peritoneal cavity and allowed to remain in contact with the peritoneum long enough for equilibrium to take place by diffusion.

QUESTION 82

Dialysis utilizes the principle of _____.

ANSWER

diffusion

QUESTION 83

Dialysis is the differential diffusion of certain particles through a _____ membrane.

ANSWER

semipermeable

QUESTION 84

Substances which are restrained by their molecular size from passing through the peritoneal dialysis membrane include serum _____ and _____.

ANSWER

proteins ... blood cells

The clinical procedure whereby the peritoneal membranes are utilized for exchange of ions in an effort to regulate total body concentration is termed _____ .

QUESTION 85

peritoneal dialysis

ANSWER

DESIGN OF DIALYSIS FLUID

ITEM 28

The composition of dialysis fluid determines the direction of diffusion. Substances will diffuse across the membrane down a concentration gradient and thus may be removed from, or added to, body fluids. For example, if the patient has accumulated an excess of potassium, the dialysis fluid contains little or no potassium, thus encouraging diffusion down a concentration gradient from body fluid into the dialysate. If sodium levels need to be held stable, the dialysate contains an amount of sodium approximating that in the body fluid, so that equal concentration across the peritoneal membranes prevents net diffusion in either direction.

In essence, then, dialysis causes unwanted excesses of electrolytes, or even drugs, to diffuse across the peritoneal surfaces and become diluted by the dialysate, which is then drained. Repeating the procedure several times relieves the extracellular fluids of undesirable build-up of potentially toxic products. At the same time, the specific design of the dialyzing fluid prevents loss of needed electrolytes from body fluids.

The concentration of specific ions in the _____ fluid determines which ions will be removed from body fluids.

QUESTION 86

dialysis

ANSWER

Specific ions go from an area of _____ concentration to an area of _____ concentration across peritoneal membranes in a direction down a concentration _____ .

QUESTION 87

high . . . low . . . gradient

ANSWER

Dialysis fluid is designed in such a way that excess ions and molecules in body fluids will diffuse into the dialysate down a _____ gradient.

QUESTION 88

concentration

ANSWER

BASIC PHYSIOLOGY OF FLUIDS AND ELECTROLYTES

QUESTION 89   The conservation of needed substances is accomplished by having the concentration of these particular substances in dialysis fluid _____ their concentration in body fluids.

ANSWER   equal

ITEM 29   WORKING EXAMPLE OF DIALYSIS

The diagram below illustrates a hypothetical case showing the effectiveness of peritoneal dialysis in a patient with renal insufficiency. The patient has difficulty removing excess hydrogen ions from the body and is in a state of acidosis. As a consequence of the acidosis, potassium is retained. High serum potassium, or hyperkalemia, causes a risk of cardiac arrhythmia. In addition to the problem of retention of potassium, the patient's excretion of urea, a by-product of protein metabolism, is impaired.

The dialysis solution introduced into the peritoneal cavity has the following concentrations:

DIALYSIS FLUID	peritoneal membrane	BODY FLUIDS
sodium: 142 mEq/L		sodium: 136 mEq/L
potassium: 0		potassium: 8 mEq/L
chloride: 114 mEq/L		chloride: 116 mEq/L
bicarbonate: 28 mEq/L		bicarbonate: 28 mEq/L
urea: 0		urea: 50 mg%

Arrows in the diagram indicate the direction in which substances will diffuse to reach a concentration equilibrium. At equilibrium, body fluid concentration of sodium will be increased, whereas extracellular potassium, chloride, and urea will be decreased. Since the dialysis fluid contains bicarbonate in a quantity equal to that in body fluid, no net diffusion will occur.

QUESTION 90   In the hypothetical case described in Item 29 and illustrated in the foregoing diagram, which substances diffuse through the peritoneal membranes in such a way as to lower plasma levels?

(a) _____

(b) _____

(c) _____

ANSWER   potassium . . . chloride . . . urea

Unneeded excesses, e.g., potassium in a hyperkalemic patient, are removed from the body by _____ into the dialysis fluid.

QUESTION 91

diffusion

ANSWER

To remove excess potassium in the hyperkalemic patient, the potassium in the dialysis fluid must be *more/less* (circle one) concentrated in the dialysis fluid than in body fluids.

QUESTION 92

less

ANSWER

TRANSPORT BY FACILITATED DIFFUSION

ITEM 30

Substances too large for diffusion through pores are transported through the cell membrane itself by *facilitated diffusion* and by *active transport*. In facilitated diffusion, substances which are too large for pores are transported down a concentration gradient across cell membranes with the help of a carrier protein. For example, glucose, with the help of insulin, is transported across muscle cells in this manner. Whereas glucose itself is insoluble in the membrane, the glucose-carrier complex is soluble and diffuses across. Glucose separates from the carrier on the inner surface of the membrane, thus freeing the carrier for further facilitated diffusion. The facilitated diffusion of glucose is impeded in the diabetic patient, who has an inadequate or ineffective serum insulin level.

Figure 6 illustrates the difference between diffusion and facilitated diffusion. The top part of the diagram shows the diffusion of oxygen from an area of high concentration (designated O_2) into an area of lesser concentration (designated o_2). The lower part of the diagram shows facilitated diffusion in which glucose diffuses from a high concentration area to a low concentration area, but does so by combining with a carrier substance at one surface of the membrane (1). The carrier-glucose complex is soluble and diffuses to the other side of the membrane (2), where the complex splits, releasing the glucose from the carrier.

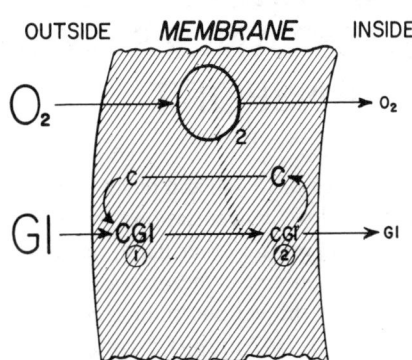

Figure 6. Diffusion of substances through the lipid matrix of the membrane. The upper part of the figure shows *free diffusion* of oxygen through the membrane and the lower part shows *carrier-mediated* or *facilitated diffusion* of glucose. (From Guyton: *Textbook of Medical Physiology.* 5th Ed. Philadelphia: W. B. Saunders Company, 1976.)

32 BASIC PHYSIOLOGY OF FLUIDS AND ELECTROLYTES

QUESTION 93 Facilitated diffusion transports substances which are too _____ for diffusion through pores.

ANSWER large

QUESTION 94 By means of facilitated diffusion, glucose is transported, with the help of _____, across muscle cell membranes.

ANSWER insulin

QUESTION 95 In the transport of glucose across muscle cell membranes, the direction of transfer is from a _____ concentration to a _____ concentration by formation of a _____ glucose-carrier complex.

ANSWER high ... low ... soluble

QUESTION 96 The process of glucose transfer by facilitated diffusion is ineffective in the _____ patient because of inadequate insulin.

ANSWER diabetic

QUESTION 97 Facilitated diffusion differs from diffusion in that facilitated diffusion requires a _____.

ANSWER carrier

ITEM 31 TRANSLOCATION OF SUBSTANCES BY ACTIVE TRANSPORT

Simple diffusion and facilitated diffusion both move substances down a gradient from high to low concentrations, with a tendency to equalize the concentrations. *Active transport*, on the other hand, maintains the distribution of certain substances against the force of extreme concentration gradients, and accounts for an unequal distribution of solutes in the various fluid compartments. The sodium level of extracellular fluid, for instance, is 142 mEq./L., compared to 10 mEq./L. in intracellular fluid. Conversely, potassium is concentrated within the cell at an intracellular value of 141 mEq./L., compared to only 5 mEq./L. in extracellular fluid.

Diffusion and facilitated diffusion move substances down a concentration _____, from an area of _____ concentration to an area of _____ concentration.

QUESTION 98

gradient ... high ... low

ANSWER

The transport mechanism which allows a substance to move against a concentration gradient is called _____.

QUESTION 99

active transport

ANSWER

Active transport accounts for an _____ distribution of substances in the various body fluids.

QUESTION 100

unequal

ANSWER

ENERGY REQUIREMENT IN ACTIVE TRANSPORT

ITEM 32

Active transport is similar to facilitated diffusion in that both need a specific carrier substance to promote solubility in the cell membrane. They differ in that active transport requires energy to overcome the concentration gradient. This energy is supplied by adenosine triphosphate stored within the cell membrane.

Figure 7 illustrates a case of active transport in which a substance moves from a low concentration, s; combines with a carrier to form the complex, CS; and moves across the

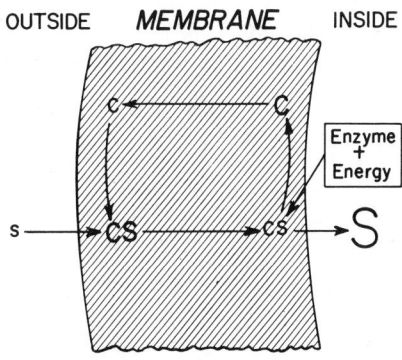

Figure 7. Basic mechanism of active transport. (From Guyton: *Textbook of Medical Physiology.* 5th Ed. Philadelphia: W. B. Saunders Company, 1976.)

membrane where energy is available for splitting the complex, releasing the substance into the cell, where a high concentration of the very same substance is already present, S. Transport against a concentration gradient is the particular characteristic of active transport. Active transport accounts for (1) absorption of glucose across the intestinal wall; (2) reabsorption of some substances from the lumina of the kidney tubules into interstitial fluids; (3) concentration of sodium outside a nerve cell and of potassium inside the cell; (4) concentration of iodine in the thyroid gland; and (5) transport of amino acids across various membranes.

QUESTION 101   Active transport is chemical combination of substance with a _____ which makes the translocated substance soluble in the cell wall.

ANSWER   carrier

QUESTION 102   Since the direction is from an area of _____ concentration to one of _____ concentration, active transport utilizes energy from the metabolic processes of the cell.

ANSWER   low . . . high

QUESTION 103   This energy is supplied by _____ _____ stored within the cell membrane.

ANSWER   adenosine triphosphate

QUESTION 104   A point of similarity between facilitied diffusion and active transport is the fact that both mechanisms utilize a _____ .

ANSWER   carrier

QUESTION 105   Active transport is responsible for reabsorption of glucose in the _____ and for the concentration of iodine in the _____ gland.

ANSWER   intestine . . . thyroid

QUESTION 106   Distribution of sodium and potassium across the nerve cell is accomplished by *facilitated diffusion/active transport* (circle one).

ANSWER   active transport

A point of difference between facilitated diffusion and active transport is that facilitated diffusion moves a substance from a _____ concentration to a _____ concentration, whereas active transport moves a substance from a _____ concentration to a _____ concentration.	QUESTION 107

high . . . low . . . low . . . high	ANSWER

THE SODIUM-POTASSIUM CARRIER IN ACTIVE TRANSPORT

ITEM 33

Figure 8 illustrates the hypothesis that a single carrier transports both sodium and potassium across a nerve cell membrane. A change in the chemical nature of the enzyme molecule makes possible the carrier's affinity for either one or the other of the two ions. When the carrier's chemical nature is such that it combines with sodium it is designated Y. When its configuration is such that it combines with potassium, the carrier is designated X. Low concentration is indicated by s; higher concentration, by S. Follow the direction of the arrows until you understand the mechanism.

Adenosine triphosphate supplies the energy for active transport. An enzyme, adenosine triphosphatase, acts on adenosine triphosphate (abbreviated ATP in the diagram) in combination with magnesium to break a high-energy phosphate bond and thus releases the energy to power the active transport system.

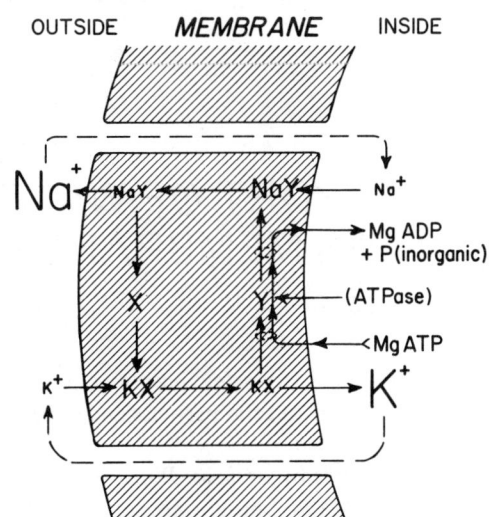

Figure 8. Postulated mechanism for active transport of sodium and potassium through the cell membrane, showing coupling of the two transport mechanisms and delivery of energy to the system at the inner surface of the membrane. (From Guyton: *Textbook of Medical Physiology*. 5th Ed. Philadelphia: W. B. Saunders Company, 1976.)

36 BASIC PHYSIOLOGY OF FLUIDS AND ELECTROLYTES

QUESTION 108   Examine the three diagrams below and circle the letter above the one which illustrates active transport.

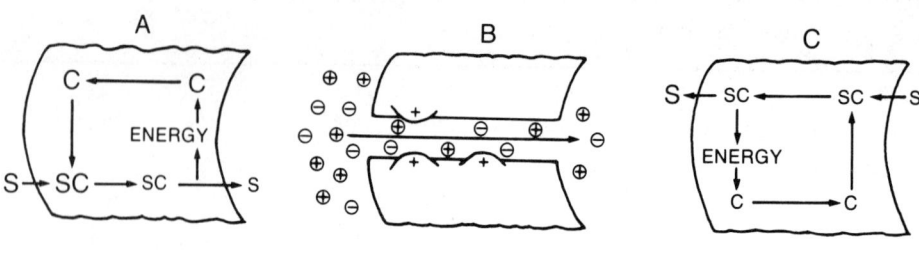

ANSWER   A. This answer is incorrect. One of the characteristics of active transport is that a substance moves into an area of higher concentration. In Diagram A, the movement is not into the high concentration area, S, but into the low concentration area, s.

B. This answer is incorrect. Diagram B illustrates diffusion downhill with a concentration gradient and an electrical gradient.

C. This answer is correct. This diagram illustrates active transport in the movement of a substance from a smaller concentration inside the membrane, s, to an area of higher concentration outside the membrane, S. The direction in this diagram goes from right to left.

QUESTION 109   Transport against a concentration gradient requires

A. a carrier

B. both a carrier and energy obtained from adenosine triphosphate

ANSWER   A. This answer is incorrect. Facilitated diffusion required a carrier only. Facilitated diffusion goes with a concentration gradient, not against it.

B. This answer is correct. Active transport requires a carrier to make the substance soluble in the membrane wall and an energy source to oppose the concentration gradient.

ITEM 34 TRANSPORT OF WATER ACROSS MEMBRANES BY OSMOSIS

The preceding Items (simple diffusion, facilitated diffusion, active transport) dealt with the transport of solid particles and gases. This Item concerns the movement of water across cell membranes.

Water diffuses by *osmosis*. When the membrane separating two fluid compartments is permeable to water but not to some of the particles dissolved in it, water will pass

Figure 9. Osmosis at a cell membrane when a sodium chloride solution is placed on one side of the membrane and water on the other side. (From Guyton: *Textbook of Medical Physiology.* 5th Ed. Philadelphia: W. B. Saunders Company, 1976.)

through the membrane down a water concentration gradient, i.e., from an area of high concentration of water into an area of low concentration of water. As this happens water moves into the compartment which has the higher concentration of solutes, or dissolved particles.

Figure 9 illustrates the diffusion of water into an area of higher osmotic pressure imposed by the larger quantity of dissolved particles. Indeed, the greater the number of dissolved particles, either molecules or ions, the greater the osmotic pressure.

QUESTION 110

Osmotic pressure is determined by the number of dissolved particles per unit volume of liquid. It is a force that causes transfer of water across a semipermeable membrane from a solution of low osmotic pressure (low solute concentration) to one of high osmotic pressure (high solute concentration). Net osmosis is transfer of water into a solution which has a

A. lower osmotic pressure

B. higher osmotic pressure

ANSWER

A. This answer is incorrect. The solution with lower osmotic pressure is the one with lower concentration of solutes (particles) and higher solvent concentration. Osmosis is toward the solution with *higher solute* concentration and *lower solvent* concentration. Read Answer B.

B. This answer is correct. Imagine two solutions separated by a membrane permeable only to water. Solution A contains 5% glucose; solution B, 10% glucose. Side A, having less solute (glucose), contains more solvent (water); side B has more glucose (solute) but less water (solvent). Since osmotic pressure varies with the concentration of solute, side B with its greater concentration of glucose is the solution with the higher osmotic pressure. Since osmosis moves water from the high solvent concentration area to the low solvent concentration area until water equilibrium is established, the osmotic movement will be from side A to side B—from the solution of low osmotic pressure into the solution of higher osmotic pressure.

SUMMARY OF UNIT ONE

Essentially all organ systems play some part in homeostasis, or maintaining the constancy of the internal environment. A system of factors which tend to alter this constancy and factors which tend to reestablish constancy results in "fluid and electrolyte balance."

Many diseases affect various organ systems and render homeostatic mechanisms ineffective. When any of these disturbances occur, cell function and consequently organ function are compromised. Often it is necessary for fluid and electrolyte balance to be clinically reestablished by intravenous therapy, dialysis procedures, or other methods, until the body itself can resume its normal role.

Body fluids are compartmentalized into intracellular fluids and extracellular fluids. These two compartments are separated by cell membranes. Intracellular and extracellular fluids differ primarily in the concentration of electrolytes. Intracellular fluid is high in potassium, phosphate, and magnesium. Extracellular fluids are high in sodium, calcium, chloride, and bicarbonate. Cellular function is dependent upon having the optimum distribution of ions both within the cell and outside the cell membrane.

Body fluids and their dissolved substances are in a constant state of mobility. There is a continual intake and output within the body as a whole and between the various compartments. Fluid easily shifts across cell membranes, following a direction dictated by hydrostatic pressure gradients and by osmotic pressures within the compartments. Other mechanisms, namely, simple diffusion, facilitated diffusion, and active transport, shift electrolytes from one membrane compartment to another. The body's maintenance of stable concentrations of electrolytes amidst this continual shifting of fluid and dissolved particles is called *dynamic equilibrium.*

Simple diffusion moves small water-soluble molecules and ions, oxygen, carbon dioxide, and other lipid-soluble substances.

Facilitated diffusion allows glucose, in the presence of insulin, to enter muscle cells; without this mechanism, the molecular size of glucose would prevent its entry through pores.

Active transport accounts for (1) absorption of glucose across the intestinal wall; (2) reabsorption of some substances from the lumina of the kidney tubules into interstitial fluids; (3) concentration of sodium on the outer surface of a nerve cell membrane and of potassium inside the nerve cell membrane; (4) concentration of iodine in the thyroid gland; and (5) transport of amino acids.

Fluid shifts occur secondary to ionic concentrations. The direction in which water will move across a cell membrane is determined by hydrostatic pressures and by osmotic pressure on each side of that membrane. Osmotic pressure, in turn, is determined by the total number of dissolved particles on either side of the membrane. Therefore, the distribution of electrolytes and molecules in any compartment is a primary determinant of the water concentration in that particular compartment.

QUESTION 111 Essentially, all organ systems play some part in _____, or maintenance of the constancy of the internal environment.

ANSWER homeostasis

QUESTION 112 Consideration of the optimum concentration of water and dissolved substances within the body as a whole and within compartments is clinically termed _____ and _____ balance.

ANSWER fluid ... electrolyte

TRANSPORT OF WATER ACROSS MEMBRANES BY OSMOSIS

Body fluids are divided into _____ fluids and _____ fluids, each with a normal distribution of _____. — **QUESTION 113**

intracellular ... extracellular ... ions — **ANSWER**

Sodium, calcium, chloride, and bicarbonate are concentrated in _____ fluid, whereas potassium, phosphate, and magnesium are concentrated in _____ fluids. — **QUESTION 114**

extracellular ... intracellular — **ANSWER**

Cellular function is dependent on the total fluid and electrolyte content of the body and upon the _____ within compartments. — **QUESTION 115**

distribution — **ANSWER**

Concentration of fluid and electrolytes within the body as a whole is the responsibility of various organ systems, particularly the _____. — **QUESTION 116**

kidney — **ANSWER**

Distribution across cell membranes is due to osmosis, diffusion, _____, and _____. — **QUESTION 117**

facilitated diffusion ... active transport — **ANSWER**

When measured clinically, electrolytes are measured in terms of _____, which are units of chemical _____. — **QUESTION 118**

milliequivalents ... reactivity — **ANSWER**

Plasma electrolyte concentrations in any given patient are reported in _____ _____ _____, are compared to normal values, and any difference is used as a basis for determining therapeutic measures. — **QUESTION 119**

milliequivalents per liter (mEq./L.) — **ANSWER**

UNIT TWO • FLUID BALANCE

COMPARTMENTALIZATION OF BODY FLUIDS

ITEM 1

Unit One of this program dealt with the chemical composition of intracellular and extracellular fluids, the concentration of electrolytes and other substances in these fluids, the measurement of electrolyte concentrations in terms of chemical reactivity, and the various membrane transport mechanisms which bring about the differing distribution of substances in the fluids.

The broad purpose of this present unit is threefold: to investigate (1) compartmentalization of fluid between intracellular and extracellular fluid space, (2) total intake and output of body fluids under normal *and* pathological conditions, and (3) the maintenance of osmotic equilibrium within the various compartments.

Figure 10 shows the compartmentalization of body water into extracellular fluid and intracellular fluid and the approximate volume relationships. About two-thirds of the total body water is within cells. The other one-third is divided among the extracellular compartments: interstitial fluid, serum, cerebrospinal fluid, intraocular fluid, and the potential fluid spaces such as the peritoneal cavity, the pleural cavity, and so on. Total body fluids in an average 70 kg. man occupy a volume of approximately 40 liters. Intracellular fluids make up about 25 liters, of which 2 liters are red blood cell volume. Extracellular fluids come to about 15 liters, 3 liters of which are serum volume.

The chemical composition of extracellular and intracellular fluids differs as to *kinds of/concentration of* (circle one) substances.

QUESTION 1

concentration of

ANSWER

Solute concentration differences occur because of membrane _____ mechanisms.

QUESTION 2

transport

ANSWER

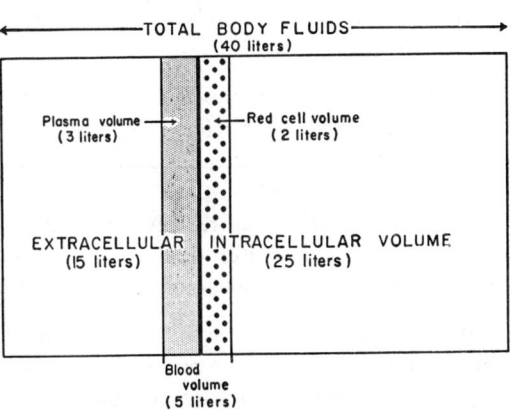

Figure 10. Diagrammatic representation of the body fluids, showing the extracellular fluid volume, intracellular fluid volume, blood volume, and total body fluids. (From Guyton: *Textbook of Medical Physiology.* 5th Ed. Philadelphia: W. B. Saunders Company, 1976.)

FLUID BALANCE

QUESTION 3 Distribution of electrolytes across membranes which separate body fluid compartments is effected by membrane _____ _____.

ANSWER transport mechanisms

QUESTION 4 The proportion of total body water within cells is about _____; the proportion in the extracellular compartment is about _____.

ANSWER two-thirds . . . one-third

ITEM 2 THE PRINCIPLE OF EQUILIBRIUM

Intake and output of varying quantities of fluids result, obviously, in a constant interchange of these fluids. Fluids also interchange between the various compartments. The state of body fluids is one of *dynamic equilibrium*. Although composition remains relatively stable, there is a continuous mixing of fluids, losses, and replacements and a never-ending turnover of molecules and ions. This dynamic equilibrium applies to water as well as to all the numerous substances dissolved in the body fluids. However, since we are concerned at the moment primarily with the *fluid* component, we will confine our exploration to total body water and fluid shifts.

Distribution of water obeys the principle of osmosis, with osmotic equilibrium maintained at all times. Osmotic equilibrium means that body fluids distribute themselves across membrane structures in such a way as to equalize osmotic pressure. Osmotic pressure, in turn, depends upon the excess of non-diffusible particles on one side of a membrane compared to the number on the other side. Although water moves in both directions across the membrane, a greater amount shifts in a direction from high water concentration (low solute) toward low water concentration (high solute). The final effect, or net movement of water, is into the area which has the greatest number of non-diffusible particles—that is, the area of higher osmotic pressure.

QUESTION 5 Dynamic equilibrium means that

A. under normal conditions the electrolyte composition of various body fluids is within a fixed stable value

B. the electrolyte composition of various body fluids stays within a relatively narrow range throughout an uninterrupted succession of losses and replacements

ANSWER A. This answer is incorrect. While the noun *equilibrium* may connote stability, the adjective *dynamic* indicates that *dynamic equilibrium* is not a fixed condition, but a relative constancy of composition obtaining throughout a constantly changing situation.

B. This answer is correct. Dynamic equilibrium means that the relatively narrow range of normal values is the result of mechanisms which regulate the concentrations of body fluids under the restless circumstances of a constant turnover of individual molecules and ions.

Osmotic equilibrium means that

A. the water component of body fluids moves across semipermeable membranes in a direction which tends to equalize osmotic pressures across these membranes

B. the electrolyte component of body fluids moves across semipermeable membranes in numbers which tend to equalize osmotic pressures across these membranes

QUESTION 6

A. This answer is correct. Shifts of fluid, specifically of the water component, follow the principle of osmosis. Remember that the net direction of flow of water across a membrane will be from the side where there are more water molecules (the dilute solution) to the side where there is less water (more solute particles).

B. This answer is incorrect. In any situation which calls for osmotic equilibrium, it is the water component which does the moving, not the electrolyte component.

ANSWER

The extent to which osmosis takes place depends on the difference in concentrations of _____ particles on each side of a membrane.

QUESTION 7

---

non-diffusible

ANSWER

HYPOTONIC, HYPERTONIC, AND ISOTONIC SOLUTIONS

ITEM 3

Osmosis is the net transfer of water across a semipermeable membrane to the side which has the higher osmotic pressure. The effect of osmosis on red blood cells is the same as its effect on any other cells of the body.

The next three Items will describe what happens to red blood cells as a result of osmosis in hypotonic, hypertonic, and isotonic solutions. A solution is *hypotonic* to intracellular fluid if it is more dilute than intracellular fluid, that is, contains fewer particles than the fluid inside the cells. A solution is *hypertonic* to intracellular fluid if it is more concentrated than intracellular fluid. An *isotonic* solution is one which has the same concentration of dissolved particles as intracellular fluids.

Osmosis is the net transfer of _____ across a semipermeable membrane into the compartment which has the larger number of _____ particles.

QUESTION 8

---

water . . . non-diffusible

ANSWER

A solution which has a concentration of dissolved particles equal to intracellular fluids is said to be _____ .

QUESTION 9

---

isotonic

ANSWER

46 FLUID BALANCE

QUESTION 10 A solution which has a concentration of dissolved particles greater than intracellular fluids is termed _____.

ANSWER hypertonic

QUESTION 11 A solution which has a concentration of dissolved particles less than that found in intracellular fluid is said to be _____.

ANSWER hypotonic

ITEM 4 FLUID SHIFTS AND RED BLOOD CELLS—HEMOLYSIS

In Figure 11, Part A shows red blood cells suspended in a hypotonic solution; that is, one of lower osmotic pressure than intracellular fluid. Obeying the principle of osmosis, water passes from this hypotonic solution across the membrane into the cell in an attempt to equalize osmotic pressure. The cells swell from the incoming water until some finally rupture and release their contents into the surrounding solution. This process is called *hemolysis*.

Obviously, the rupture and release of cell contents increases the concentration of particles in the surrounding solution. Red cells continue to hemolyze only until enough cells have added their contents to the serum to balance the concentrations on both sides of the cell membranes. Thus a slight degree of hemolysis in the body is self-limiting and

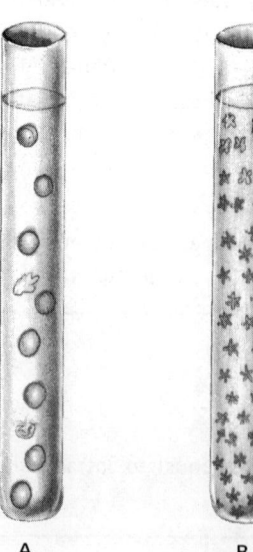

Figure 11. Hemolysis and crenation. *A*, Blood cells in hypotonic solution. Cells have become swollen and some have already burst, or hemolyzed. *B*, Blood cells in hypertonic solution. Note shriveled, or crenated, appearance. (From Dienhart: *Basic Human Anatomy and Physiology.* 2nd Ed. Philadelphia: W. B. Saunders Company, 1973.)

ceases when a balance of osmotic pressure across the membrane is achieved. Extensive intravascular hemolysis consequent upon an abnormally dilute state of serum, on the other hand, is a serious pathological condition.

When red cells are placed in a solution which is more dilute than intracellular fluid, water shifts from the _____ compartment into the _____ compartment.

QUESTION 12

extracellular . . . intracellular

ANSWER

A solution which contains a fewer number of particles than is found in body fluids is said to be _____.

QUESTION 13

hypotonic

ANSWER

If red cells are suspended in a dilute solution, _____ occurs until rupture of the red cells restores the equality of concentration inside and outside the cells.

QUESTION 14

osmosis

ANSWER

Osmosis and hemolysis continue until rupture of red cells and release of contents have increased the concentration of _____ fluids.

QUESTION 15

extracellular

ANSWER

Extensive intravascular hemolysis is a serious pathological condition which could result from an abnormally _____ state of blood serum.

QUESTION 16

hypotonic (dilute)

ANSWER

FLUID SHIFTS AND RED BLOOD CELLS—CRENATION

ITEM 5

A hypertonic solution contains a greater number of particles per liter than does intracellular fluid. Part B of Figure 11 illustrates the distribution of particles across the cell membrane when red cells are suspended in a hypertonic solution. The movement of water is now in the reverse direction from that in the case of the hypotonic solution. It

FLUID BALANCE

shifts from the interior of the cell across the cell membrane into the surrounding solution. This shift will continue until the water and solute concentrations are equal on both sides of the cell membrane. In the process, cell membranes shrink to accommodate to the lesser volume of water within the red cells. The notched appearance resulting from shrinking of red blood cells to adjust to loss of water is called *crenation*.

In severe cases of dehydration due to water deficit, a similar situation could occur. Consider also why shipwreck victims in a lifeboat are cautioned against drinking hypertonic ocean water.

QUESTION 17 When red cells are suspended in saline at a concentration which causes fluid to shift from the interior of the cell into the extracellular fluid, these cells _____ to accommodate the lesser fluid volume.

ANSWER shrink

QUESTION 18 This shriveled appearance of red blood cells is a condition called _____.

ANSWER crenation

QUESTION 19 Fluid which is more concentrated than a reference fluid is said to be *hypotonic/hypertonic* (circle one).

ANSWER hypertonic

QUESTION 20 When red blood cells are suspended in a solution which is more concentrated than intracellular fluids, osmosis causes water to shift from _____ space to _____ space.

ANSWER intracellular . . . extracellular

QUESTION 21 Fluid shifts across a semipermeable membrane until the concentration of solute and solvent are equal to establish osmotic _____.

ANSWER equilibrium

RED BLOOD CELLS IN ISOTONIC SOLUTION

ITEM 6

An isotonic solution is one in which the concentration of particles in extracellular fluid is equal to that of intracellular fluid. A solution with a concentration of 0.9 gram of sodium chloride per 100 milliliters of solution is referred to variously as *isotonic saline*, *0.9% saline*, or *normal saline*. Normal saline contains approximately the same number of particles in solution as that found in intracellular fluid.

A solution of 0.9% saline is said to be _____ with reference to intracellular fluid.

QUESTION 22

---

isotonic

ANSWER

A solution of 0.9% saline is also referred to as _____ saline.

QUESTION 23

---

normal (or isotonic)

ANSWER

The number of particles in 0.9% saline is equal to the number of particles of total solute within _____.

QUESTION 24

---

cells

ANSWER

If red blood cells are suspended in 0.9% saline, the net shift of water will be _____.

QUESTION 25

---

zero

ANSWER

Red cells suspended in isotonic saline are in a solution in which osmotic _____ already exists.

QUESTION 26

---

equilibrium

ANSWER

ITEM 7 — WATER BALANCE

Water balance is important because of its involvement in the functioning of essentially all organ systems. As a transporting substance, water maintains the volume of the vascular system, provides the fluid medium for transferring nutrients and for the elimination of wastes, and aids in regulation of body temperature. Optimal water concentration within cells is necessary for many enzymatic reactions and other metabolic processes, as well as for hydrolysis of digestive products.

Normally, fluid balance is maintained when intake of water approximates output and when compartmentalization of fluids is within the normal range. Day-to-day variations of intake are compensated for by the kidneys' excretion of dilute or concentrated urine, as the situation demands. Compartmentalization is regulated primarily by osmotic fluid shifts.

QUESTION 27 In elimination of wastes and transfer of nutrients, water serves as a _____ _____.

ANSWER fluid medium

QUESTION 28 Water is essential to functioning of the cardiovascular system in maintaining blood _____.

ANSWER volume

QUESTION 29 As a cooling agent, water plays an important role in regulation of _____ _____.

ANSWER body temperature

QUESTION 30 Numerous metabolic processes and enzymatic reactions depend on _____ of water within _____.

ANSWER concentration . . . cells

QUESTION 31 Larger than usual daily water intake is normally compensated for by excretion of _____ urine; less than usual daily intake, by excretion of _____ urine.

ANSWER dilute . . . concentrated

FLUID INTAKE   ITEM 8

Water enters the body by way of beverages and liquid-containing foods, and as water released through catabolic breakdown of foodstuffs. Approximately 2600 ml. of fluid enter the body each day by these routes.

Many clinical conditions create an excess or deficiency of water in the body, placing a strain on the mechanisms of water balance. Intake of water is insufficient, for example, in comatose patients. Water intake can also be reduced in patients who are too old, too young, or too sick to satisfy the demands of thirst. In a hospital situation, fluid intake is often controlled by intraveous infusion. The nature of the infusion may vary, depending on whether the patient has an imbalance which must be corrected or simply a normal fluid balance which must be maintained.

In a clinical condition involving insufficient fluid intake, water balance must often be maintained, or imbalance corrected, by _____.   QUESTION 32

intravenous infusion   ANSWER

Water is taken into the body as fluids and food which enter the _____ system.   QUESTION 33

gastrointestinal   ANSWER

In the comatose patient, water deficit would ordinarily be attributable to   QUESTION 34

A. water loss from dehydration

B. insufficient fluid intake

The correct answer is B, insufficient fluid intake.   ANSWER

FLUID OUTPUT   ITEM 9

An average of 1500 ml. of water is removed from the body each day in urine. Approximately 100 ml. is lost in feces. These "sensible" water losses can vary, of course, with variations in water intake and abnormal gastrointestinal activity.

"Insensible" water loss occurs through the skin and lungs. Normally about 1000 ml. per day, water loss through skin increases tremendously, and becomes "sensible," when

52 FLUID BALANCE

fever causes profuse diaphoresis, or excess perspiration. Excess water loss may also occur through respiration with prolonged fever and hyperventilation. Other conditions occasioning water loss are prolonged vomiting and diarrhea and loss through burned areas of the skin. Water is *retained* in congestive heart failure and in certain kidney diseases.

QUESTION 35
The major portion of "sensible" water loss occurs with excretion of _____, while most "insensible" water loss occurs through the _____ and the _____.

ANSWER
urine . . . skin . . . lungs

QUESTION 36
Fever increases water loss through the _____ and _____.

ANSWER
lungs . . . skin

QUESTION 37
Congestive heart failure is accompanied by water *loss/retention* (circle one).

ANSWER
retention

QUESTION 38
In addition to abnormal water loss with fever, three other conditions accompanied by abnormal water loss include _____, _____, and _____.

ANSWER
vomiting . . . diarrhea . . . burns (in any order)

ITEM 10 REMOVAL AND CONSERVATION OF BODY FLUID

When total body water is excessive, the kidneys remove the excess by increased urinary output, or *diuresis*. If there is a deficiency of body fluid, the kidneys conserve water by decreasing urinary output, or *antidiuresis*. Diuresis implies a large dilute quantity of urine, while antidiuresis implies a small, concentrated quantity of urine.

A hormone secreted by the posterior pituitary gland acts on the kidneys to cause them to conserve water; the action of this hormone is

QUESTION 39

A. diuretic

B. antidiuretic

A. This answer is incorrect. Diuresis means removal of water by increased urinary volume, not retention of water.

B. This answer is correct. The hormone referred to is called *antidiuretic hormone* because it effects antidiuresis, or reabsorption of water by the kidneys.

ANSWER

The urine excreted under the influence of antidiuretic hormone is *dilute/concentrated* (circle one).

QUESTION 40

---

concentrated

ANSWER

ANATOMY AND PHYSIOLOGY OF THE KIDNEY

ITEM 11

A brief review of the structure and function of the kidney will aid in understanding the processes whereby this organ removes or retains fluid to meet total body water needs.

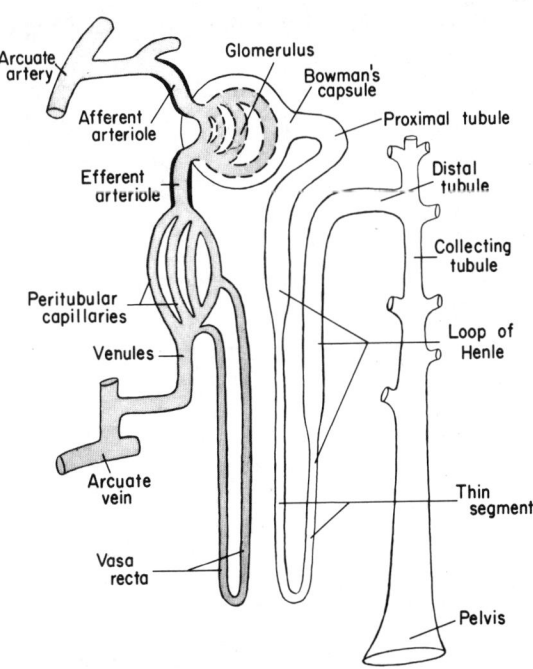

Figure 12. The functional nephron. (From Guyton: *Textbook of Medical Physiology.* 5th Ed. Philadelphia: W. B. Saunders Company, 1976.)

54   FLUID BALANCE

Each kidney is made up of several million *nephrons*, each of which has a similar structure. Essentially, a nephron is composed of a glomerulus and a tubule, as depicted in Figure 12.

The *glomerulus*, a capsule-like structure enclosing a tuft of capillaries, filters water and other substances from blood. The filtered fraction is termed *glomerular filtrate*.

The *tubule* comprises the following functionally distinct segments: the *proximal convoluted tubule*, the *loop of Henle*, the *distal convoluted tubule*, and a *collecting duct*. Glomerular filtrate flows from the glomerulus through the successive segments of the tubule. Some substances are reabsorbed from glomerular filtrate into blood through the *peritubular capillary plexus*, a network of capillaries which surround the tubule. Other substances are secreted from the peritubular capillaries into glomerular filtrate. In the collecting duct, final modification of glomerular filtrate takes place before it is excreted as urine.

QUESTION 41   The unit structure of the kidney is the _____ .

ANSWER   nephron

QUESTION 42   Separation of filtrate from serum occurs across the membranes of the _____ .

ANSWER   glomerulus

QUESTION 43   Reabsorption and secretion take place along the length of the _____ of the nephron.

ANSWER   tubule

QUESTION 44   The segment of the nephron where fluid and particles filtered from blood undergo final modification to become urine is called the _____ _____ .

ANSWER   collecting duct

QUESTION 45   The fluid which is separated from blood by the process of filtration is called _____ .

ANSWER   glomerular filtrate

The network of capillaries surrounding the tubule is termed the _____ _____ plexus.

QUESTION 46

peritubular capillary

ANSWER

The route of glomerular filtrate is through (1) the _____ convoluted tubule, (2) the _____ of _____, and (3) the _____ convoluted tubule, and from there into the collecting duct.

QUESTION 47

proximal ... loop ... Henle ... distal

ANSWER

RENAL BLOOD FLOW   ITEM 12

Examine Figure 12 again as you read through this Item. Blood flows from the *arcuate artery*, enters the glomerulus through the *afferent arteriole*, and leaves by way of the *efferent arteriole*. Within the capsule, blood cells and serum proteins are restrained from crossing the glomerular membrane because of their size. However, smaller particles, including electrolytes, amino acids, glucose, and metabolic wastes, as well as large quantities of water, are filtered from blood to form glomerular filtrate. After blood flows through the peritubular capillaries, where reabsorption from and secretion into glomerular filtrate takes place, it returns to venous circulation.

Blood enters the glomerulus by way of the _____ arteriole and leaves by way of the _____ arteriole.

QUESTION 48

afferent ... efferent

ANSWER

Glomerular filtrate is composed of water, electrolytes, and other particles which filter out of serum across the _____ _____.

QUESTION 49

glomerular membrane

ANSWER

Substances which do not filter out of blood across the glomerular membrane include _____ _____ and serum _____.

QUESTION 50

blood cells ... proteins

ANSWER

ITEM 13   FILTRATION PRESSURE

The balance of various pressures determines the volume and rate of filtration across the glomerular membrane in a manner similar to the action of forces which determine flow across a capillary membrane. Figure 13 illustrates the balance of various pressures. Note in Figure 13 the values in different parts of the nephron: mean blood pressure in the afferent arteriole is 100 mm. Hg; blood hydrostatic pressure in the glomerular capillary bed, 60 mm. Hg; pressure in the efferent arteriole, 18 mm. Hg; pressure of glomerular filtrate in the tubule, 18 mm. Hg. The arrows in Figure 13 indicate that hydrostatic pressure in the glomerular capillary bed promotes filtration, while pressures in the efferent arteriole and in the tubule oppose it.

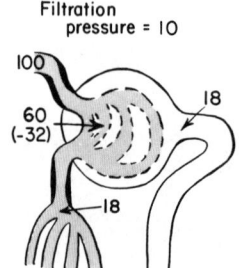

Figure 13. Normal pressures at different points in the nephron, and the normal filtration pressure. (From Guyton: *Textbook of Medical Physiology*. 5th Ed. Philadelphia: W. B. Saunders Company, 1976.)

Since capillary hydrostatic pressure (60 mm. Hg) is higher than the opposing pressures (18 mm. Hg and 18 mm. Hg), the net effect favors filtration.

QUESTION 51   Glomerular filtration takes place because of pressure _____.

ANSWER   gradients (differences)

QUESTION 52   The principal factor which promotes glomercular filtration is capillary _____ pressure, which is normally about _____ mm. Hg.

ANSWER   hydrostatic (blood) ... 60

ITEM 14   PRESSURE RELATIONSHIPS IN THE NEPHRON

Two factors oppose filtration across the glomerular membrane: serum colloid osmotic pressure and the pressure of glomerular filtrate in the efferent arteriole. The colloid osmotic pressure of approximately 32 mm. Hg is due to the presence of plasma proteins, principally albumin. Osmotic forces created by the proteins in blood retard filtration;

that is, they tend to keep fluid from leaving blood. The pressure of glomerular filtrate on the downstream side of the membrane, about 18 mm. Hg, likewise retards filtration of fluid through the glomerular membrane.

Figure 13 illustrates the pressure relationships of the factors which favor filtration and those which oppose it. A capillary hydrostatic pressure of 60 mm. Hg favors filtration. The combined total of serum colloid osmotic pressure, 32 mm. Hg, and the pressure of glomerular filtrate, 18 mm. Hg, is 32 + 18, or 50. The difference between the pressure of 60 mm. Hg favoring filtration and the restraining force of 50 mm. Hg is a net pressure of 10 mm. Hg promoting movement of fluid from the glomerular capillaries into the tubular complex.

Serum colloid osmotic pressure, 32 mm. Hg, and the pressure of glomerular filtrate, 18 mm. Hg, *promote/retard* (circle one) filtration.

QUESTION 53

---

retard

ANSWER

The sum total of pressure factors which oppose filtration is _____ mm. Hg.

QUESTION 54

50

ANSWER

Net filtration pressure is the difference between _____ pressure on the one hand and the combined total of _____ pressure and _____ pressure on the other.

QUESTION 55

---

capillary hydrostatic ... serum colloid ... glomerular filtrate

ANSWER

ARTERIOLAR CONSTRICTION AND FILTRATION RATE

ITEM 15

Constriction of the afferent arteriole lowers glomerular filtration rate by reducing the hydrostatic pressure in the glomerular capillary plexus. With a normal net filtration pressure of 10 mm. Hg favoring outward flow, capillary hydrostatic pressure could, theoretically, be reduced by this amount before filtration stops and anuria ensues.

Strong sympathetic stimulation, as occurs with massive hemorrhage, constricts afferent arterioles. Hypovolemia resulting from hemorrhage activates the sympathetic nervous system as a means of maintaining blood pressure. The oliguria or anuria consequent upon arteriolar constriction in such a case tends to conserve extracellular fluid volume as a means of correcting the hypovolemia.

58 FLUID BALANCE

QUESTION 56
Capillary hydrostatic pressure is reduced by strong _____ stimulation, which constricts the _____ _____ through which blood enters the glomerular capillary bed.

ANSWER
sympathetic ... afferent arterioles

QUESTION 57
Following hemorrhage, glomerular filtration rate decreases because of a decrease in _____ _____ pressure accompanying sympathetic stimulation.

ANSWER
capillary hydrostatic (blood)

QUESTION 58
A reduction in capillary blood pressure reduces _____ _____ rate and thereby leads to a condition of _____.

ANSWER
glomerular filtration ... oliguria

ITEM 16 COLLOID OSMOTIC PRESSURE AND FILTRATION RATE

Changes in serum colloid osmotic pressure affect the glomerular filtration rate. For example, consumption of large quantities of liquid lowers serum osmotic pressure by hemodilution. Because the serum osmotic pressure which opposes filtration is reduced, filtration takes place more freely, leading to increased urinary output. Thus the net effect of hydration is polyuria. In the case of dehydration, the action of the foregoing mechanisms is exactly reversed.

QUESTION 59
Glomerular filtration rate varies with changes in serum _____ _____ pressure.

ANSWER
colloid osmotic

QUESTION 60
Hydration increases glomerular filtration rate because it lowers osmotic pressure by dilution of _____.

ANSWER
blood

An increased glomerular filtration rate due to hydration _____ urinary output; the net effect is _____.

QUESTION 61

increases ... polyuria

ANSWER

CHRONIC GLOMERULAR NEPHRITIS AND FILTRATION RATE

ITEM 17

Chronic glomerular nephritis is among the pathological conditions which alter the pressure relationships across the glomerular membrane. Glomerular nephritis is characterized by inflammation of the glomerulus and a marked increase in the permeability of the glomerular membrane. If the condition is sufficiently aggravated, the membrane can no longer restrain protein substances, particularly albumin, from passing out of blood through the membrane pores into the glomerular filtrate. With a depletion of serum protein, the osmotic forces which retard filtration are diminished. At the same time that osmotic restraining forces are diminished by serum protein loss, the osmotic pressure of glomerular filtrate is intensified because of the presence of albumin. Thus glomerular nephritis brings on acceleration of filtration rate both by reducing opposing pressure and by increasing favorable pressure. The increased glomerular filtration rate produces a tendency toward greater urinary output, or polyuria.

Glomerular nephritis results in *a decrease/an increase* (circle one) in the permeability of the glomerular membrane.

QUESTION 62

an increase

ANSWER

As protein leaks across the glomerular membrane, osmotic pressure of serum *increases/decreases* (circle one) while the osmotic pressure of glomerular filtrate *increases/decreases* (circle one).

QUESTION 63

decreases ... increases

ANSWER

The shift of osmotic pressures in nephritis tends to *accelerate/retard* (circle one) glomerular filtration rate and lead to *oliguria/polyuria* (circle one).

QUESTION 64

accelerate ... polyuria

ANSWER

60 FLUID BALANCE

ITEM 18 THE ANTIDIURETIC HORMONE MECHANISM

Two hormones, antidiuretic hormone and aldosterone, are primarily responsible for fluid balance. The first, *antidiuretic hormone*, abbreviated ADH, is secreted by the posterior pituitary gland. As its name implies, antidiuretic hormone suppresses *diuresis*, or the secretion of urine. Specifically, in the case of low fluid intake, the kidneys, through the mediation of the ADH mechanism, conserve body water by forming a small quantity of concentrated urine. On the other hand, when fluid intake is excessive, the posterior pituitary gland secretes less ADH, and the kidneys remove the excess fluid by forming a large quantity of dilute urine. Urinary volume can vary from as little as 400 milliliters per day, when extreme quantities of ADH are secreted, to as much as 20 liters per day, when ADH production diminishes.

QUESTION 65   The role of ADH in regulating the water concentration of urine is to increase the kidneys' *reabsorption/excretion* (circle one) of water.

ANSWER   reabsorption

QUESTION 66   A patient with diabetes insipidus secretes little or no ADH. His urinary volume is

A. low

B. high

ANSWER   A. This answer is incorrect. ADH secretion promotes antiduresis, or retention of body water. The diabetic patient in this case does not produce ADH, does not retain water, and therefore has a high urinary volume. Read Answer 66–B.

B. This answer is correct. Since the ADH supply is inadequate, water is not being retained but is excreted in a high volume of urine.

ITEM 19 OSMORECEPTORS AND SECRETION OF ADH

Figure 14 illustrates the mechanism which controls the secretion of antidiuretic hormone. The secretion of ADH is triggered by a discharge of nerve impulses from the hypothalamus. Neurons in the hypothalamus, called *osmoreceptors*, respond to the solute concentration of fluid surrounding them. When the body intake of water is low, the solute concentration of the fluid perfusing the hypothalamic area increases. To achieve equilibrium, water moves by osmosis from within the osmoreceptor cells into the surrounding extracellular fluid. This loss of intracellular fluid causes the osmoreceptors to

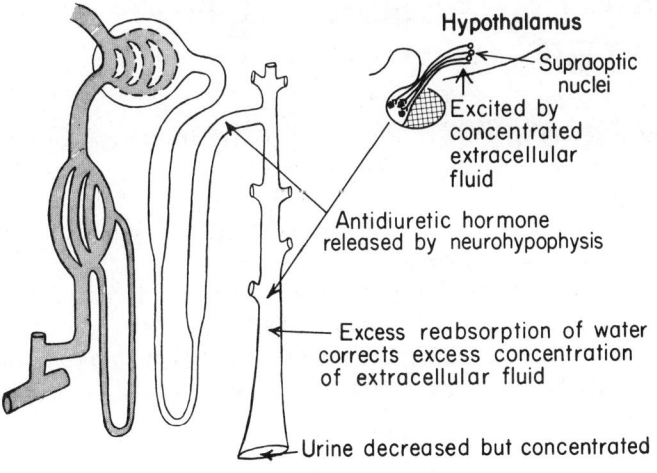

Figure 14. Control of extracellular fluid osmolality by the osmoreceptor system. (From Guyton: *Textbook of Medical Physiology.* 4th Ed. Philadelphia, W. B. Saunders Company, 1971.)

shrink, and the shrinkage excites the hypothalamus to increase its rate of discharge of nerve impulses to the posterior pituitary. With increased frequency of impulses from the hypothalamus, the posterior pituitary releases greater quantities of ADH into the circulatory system.

The osmoreceptor mechanism works in the following manner:

(1) Osmoreceptor neurons respond to *increased/decreased* (circle one) solute concentration of fluid perfusing the hypothalamus.

(2) *Hypotonic/hypertonic* (circle one) extracellular fluids bring on osmotic shifts which cause osmoreceptors to shrink.

(3) Shrinkage of osmoreceptor cells *increases/decreases* (circle one) the rate of nerve impulses from the hypothalamus to the posterior pituitary.

(4) Incoming nerve impulses cause the posterior pituitary to release a *smaller/greater* (circle one) quantity of ADH into the circulation.

(5) Increased circulating-levels of ADH *promote/suppress* (circle one) diuresis.

QUESTION 67

---

(1) increased

(2) hypertonic

(3) increases

(4) greater

(5) suppress

ANSWER

ITEM 20 ADH AND REABSORPTION OF WATER

ADH circulates in blood to the kidneys, where it acts on the distal tubules and collecting ducts to increase their permeability and thus bring about reabsorption of water. Water is moved from the glomerular filtrate in the lumina of the tubules to interstitial fluid and reabsorbed into peritubular capillaries. The end result is the elaboration of a small quantity of concentrated urine.

Reabsorbed water returns by way of the renal veins to the circulatory system, where it dilutes the concentration of extracellular fluids. The more dilute extracellular fluid perfusing the hypothalamus diminishes the original stimulus which activated the ADH mechanism. As ADH production slows, the kidneys excrete a larger and more dilute quantity of urine.

QUESTION 68 With renal reabsorption of water under the influence of ADH, urinary volume is *increased/decreased* (circle one), while urinary concentration is *increased/decreased* (circle one).

ANSWER decreased ... increased

QUESTION 69 Water reabsorbed from the kidney tubules into venous circulation brings about a _____ of the concentration of extracellular fluids, thus removing the original stimulus that had triggered the secretion of _____.

ANSWER dilution ... antidiuretic hormone (ADH)

ITEM 21 DIABETES INSIPIDUS AND ADH DEFICIT

Diabetes insipidus, radiation therapy, and certain infectious processes are conditions which may impair ADH secretion. In the ADH impairment which accompanies diabetes insipidus, the kidneys fail to reabsorb sufficient water from glomerular filtrate into the blood. The patient excretes a large volume of dilute urine, creating a water deficit and an intense thirst. Fluid intake in the diabetic is usually geared to replacement of that volume of fluid which is lost.

ADH secretion may be deficient following certain infectious processes and following radiation therapy because of insult to the posterior lobe of the pituitary gland. The extent of the injury determines the degree of return to normalcy. For therapy, the patient receives vasopressin (Pitressin) as an antidiuretic replacement and is given intravenous solution at a rate calculated to replace fluids lost through excessive urinary output.

A disease which has as its uniform cause a deficiency of antidiuretic hormone is _____.	QUESTION 70
diabetes insipidus	ANSWER
Damage to the _____ _____ gland following _____ therapy may impair secretion of ADH in varying degrees.	QUESTION 71
posterior pituitary ... radiation	ANSWER
Treatment of patients with an impaired ADH mechanism is aimed at replacing _____ hormone with _____ and correcting _____ of fluid with increased fluid _____.	QUESTION 72
antidiuretic ... vasopressin (Pitressin) ... loss (output) ... intake	ANSWER

Patients with diminished secretion of ADH excrete

A. small quantities of concentrated urine
B. small quantities of dilute urine
C. large quantities of dilute urine
D. large quantities of concentrated urine

QUESTION 73

The correct answer is C: large quantities of dilute urine.

ANSWER

ALDOSTERONE SECRETION AND FLUID RETENTION

ITEM 22

Aldosterone, a hormone secreted by the adrenal cortex, acts on the kidney to cause reabsorption of sodium from glomerular filtrate into blood. Increased sodium reabsorption raises the solute concentration of body fluids, and the ADH mechanism brings about the retention of water. The reabsorbed water, in turn, dilutes the hyperosmolarity induced by sodium retention. Clinically, sodium balance is invariably associated with water balance. The net result, therefore, is an increase in isotonic extracellular fluid volume.

Retention of _____ invariably accompanies reabsorption of sodium.	QUESTION 74
water	ANSWER

FLUID BALANCE

QUESTION 75   When extracellular fluid has a high sodium concentration, reabsorption of water tends to return that fluid to *a hypertonic/an isotonic* (circle one) state.

ANSWER   an isotonic

QUESTION 76   When the _____ _____ releases aldosterone, this hormone brings about reabsorption of _____ by the kidney.

ANSWER   adrenal cortex ... sodium

ITEM 23 — SYMPTOMS OF EXTRACELLULAR FLUID VOLUME EXCESS

The increase of isotonic extracellular fluid volume spoken of in the last Item may become excessive as a result of various pathological conditions. Patients with congestive heart failure, chronic kidney failure, or liver disease, or those who simply have received excessive isotonic intravenous therapy, are subject to isotonic extracellular fluid excess.

Signs and symptoms of a state of excessive ECF volume include edematous states. Systemic edema may develop because of increased capillary pressure accompanying hypervolemia; pulmonary edema, marked by dyspnea, cough, and frothy sputum, may be observed after leakage of fluid into alveoli. Some patients may show weight gains because of the retention of sodium and water. In the case of excessive isotonic ECF volume, laboratory findings show normal serum sodium values because of the retention of both water and sodium.

Treatment of ECF excess involves restriction of sodium and possibly of water, together with administration of diuretics.

QUESTION 77   Isotonic fluid excess can result from compensatory mechanisms in patients with _____ _____ failure, _____ failure, or _____ disease.

ANSWER   congestive heart ... kidney ... liver

QUESTION 78   In some cases of ECF excess, there is leakage of fluid from capillaries in the lungs, resulting in _____ edema.

ANSWER   pulmonary

Increased capillary pressure accompanying hypervolemia will bring on _____ edema.	QUESTION 79
systemic	ANSWER
Symptoms of pulmonary edema include _____, _____, and frothy _____.	QUESTION 80
dyspnea ... cough ... sputum	ANSWER
Retention of fluid in ECF excess may be indicated by the patient's _____ gain.	QUESTION 81
weight	ANSWER
In a patient with excessive isotonic ECF, laboratory serum sodium levels are _____.	QUESTION 82
normal	ANSWER
Treatment for ECF volume excess includes _____ restriction and use of _____.	QUESTION 83
sodium ... diuretics	ANSWER

SYMPTOMS OF EXTRACELLULAR FLUID VOLUME DEPLETION

ITEM 24

A situation opposite to ECF excess obtains when there is a loss of isotonic extracellular fluid, as can occur with hemorrhage, diarrhea, burns, fever, vomiting, or fistulas. In such instances, water and electrolytes are lost proportionately.

Signs and symptoms of initial extracellular fluid depletion include nausea, vomiting, dryness of the mucous membranes, and weight loss. ECF depletion may be detected from symptoms associated with hypovolemia, such as postural decrease in systolic blood pressure in excess of 10 mm. Hg or oliguria progressing to anuria as a result of decreased renal blood flow. Shock brought on by hypovolemia and circulatory collapse is also a sign of extracellular fluid depletion. In patients with ECF depletion, laboratory tests will show elevated hematocrit due to hemoconcentration, while serum sodium will appear normal, since sodium and water are lost proportionately.

If fluid depletion is severe, intravenous saline infusion may be called for:

QUESTION 84

Some of the conditions which result in isotonic ECF deficit include:

(1) _____
(2) _____
(3) _____
(4) _____
(5) _____
(6) _____

ANSWER

vomiting ... diarrhea ... burns ... hemorrhage ... fistulas ... fever (any order)

QUESTION 85

Depletion of sodium and water results in _____ fluid loss, nausea, _____ , and dryness of the _____ membranes.

ANSWER

extracellular ... vomiting ... mucous

QUESTION 86

Hypovolemia results in an excessive postural decrease in _____ blood pressure.

ANSWER

systolic

QUESTION 87

Decreased renal blood flow results in _____ progressing to _____ .

ANSWER

oliguria ... anuria

QUESTION 88

Hypovolemia and circulatory collapse can lead to _____ .

ANSWER

shock

Laboratory values in the patient with ECF depletion are elevated for _____, due to hemoconcentration, and _____ for serum sodium levels.

QUESTION 89

---

hematocrit ... normal

ANSWER

DEHYDRATION

ITEM 25

A typical clinical case of fluid imbalance is that of the elderly debilitated patient with a decreased fluid intake. Because of weakness and apathy, this patient is simply not drinking enough fluids and consequently becomes dehydrated. With decreased fluid intake, several mechanisms go into operation to conserve body water. As body fluids become more concentrated, the hypertonicity stimulates osmoreceptors, ADH secretion increases, renal reabsorption of fluid increases, and urinary volume decreases. Since solutes continue to be excreted, the patient excretes a darker colored urine. Indeed, urine color can be used as a quick rough estimate of the state of hydration of a patient.

Dehydration due to reduction in fluid intake results in *hypotonicity/hypertonicity* (circle one) of body fluids.

QUESTION 90

---

hypertonicity

ANSWER

Solute concentration in the dehydrated patient *increases/decreases* (circle one) ADH secretion and *increases/decreases* (circle one) urinary volume.

QUESTION 91

---

increases ... decreases

ANSWER

The extent of dehydration of a given patient can be roughly estimated from the _____ of the _____.

QUESTION 92

---

color ... urine

ANSWER

ITEM 26 — SYMPTOMS OF DEHYDRATION

Among the signs and symptoms that are noted in the dehydrated patient are increased blood viscosity, lowered blood pressure, and lowered central venous pressure due to hemoconcentration. Laboratory findings show elevated hemoglobin and hematocrit, with serum sodium levels greater than 150 mEq./L., as contrasted with a normal range of 132 to 142 mEq./L.

Lack of replacement of water loss results in weight loss. The patient is oliguric and excretes a concentrated urine with a specific gravity greater than 1.030; the condition may progress to anuria. Other symptoms of dehydration are dryness of the skin and mucous membranes, thirst, fever, and mental agitation and depression due to dehydration of brain cells, which, if severe enough, brings on coma.

QUESTION 93 — Hemoconcentration results in increased blood _____, decreased blood _____, and decreased central _____ pressure.

ANSWER — viscosity ... pressure ... venous

QUESTION 94 — The dehydrated patient's skin and mucous membranes are _____.

ANSWER — dry

QUESTION 95 — The mental agitation and depression of the dehydrated patient may culminate in _____.

ANSWER — coma

QUESTION 96 — Temperature in the dehydrated patient is *elevated/lowered* (circle one).

ANSWER — elevated

QUESTION 97 — In cases of dehydration, laboratory tests show increased _____ of urine, elevated _____, and elevated serum _____ levels.

ANSWER — specific gravity ... hematocrit ... sodium

The dehydrated patient tends to be *polyuric/oliguric* (circle one).

QUESTION 98

oliguric

ANSWER

FLUID VOLUME AND CONCENTRATION IN THE DEHYDRATED PATIENT

ITEM 27

Figure 15 illustrates (1) the intracellular and extracellular changes which might typically occur in a dehydrated patient and (2) the relationship between volume of fluid

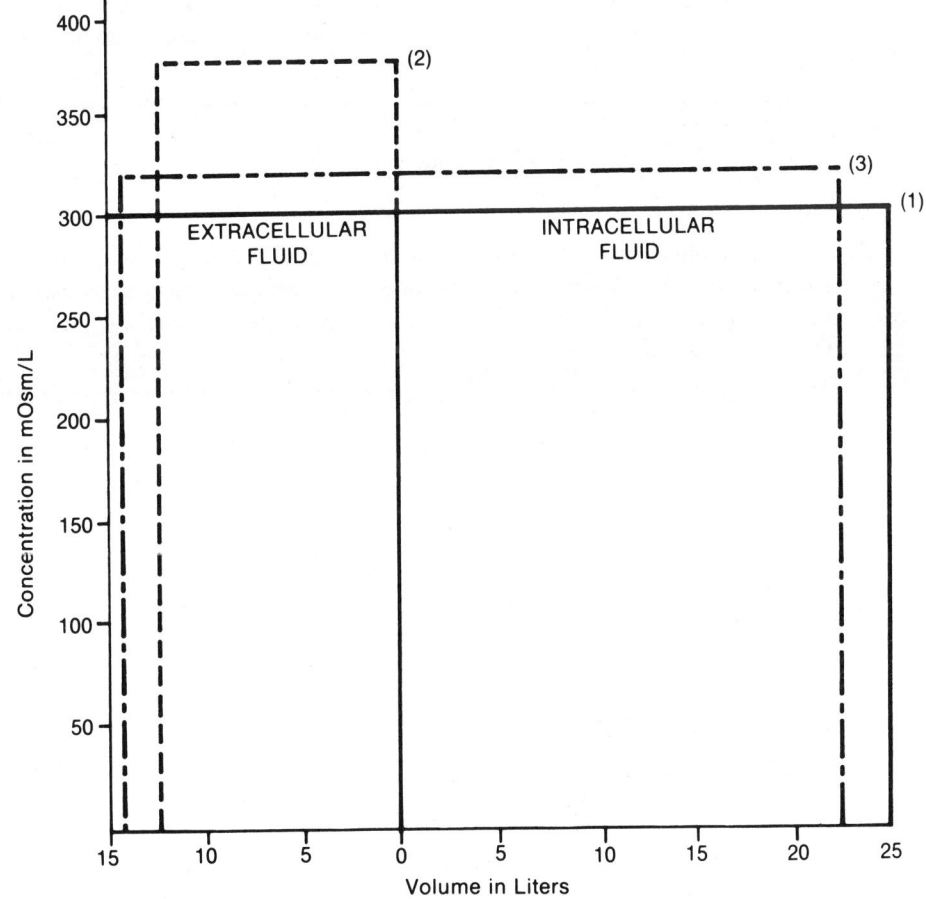

Figure 15.
(1) Solid line (———), frame #1, represents original condition of normal volume and concentration distribution.
(2) Dashed line (------), frame #2, represents the initial effect of reduced intake of water.
(3) Dot-dash line (– · –), frame #3, represents final condition after osmotic equilibrium has been achieved.

and concentration of solute particles in intracellular and extracellular compartments in such a patient.

In Figure 15, the horizontal scale represents volume in liters; the vertical scale represents concentration, or number of particles per liter, in milliosmols (mOsm./L.).

Frame #1 represents the initial condition prior to the patient's dehydration, with the normal ratio of 15 liters of extracellular fluid to 25 liters of intracellular fluid, and with solute concentration equal in both compartments at 300 mOsm./L.

Frame #2 shows the initial effect of the decreased water intake. Extracellular volume has decreased from 15 to 12 liters, as shown on the horizontal scale; extracellular solute concentration has increased from 300 to 375 mOsm./L., as shown on the vertical scale.

Frame #3 presents the final situation after osmotic equilibrium has been reestablished by movement of water from the intracellular compartment into the extracellular compartment. Solute concentration has been equalized in both compartments at 325 mOsm./L. Water deficit is shared by both compartments, with 14 liters as the extracellular volume and 23 liters as the intracellular volume.

QUESTION 99 In the dehydrated patient, the original deficit manifests itself as a decrease in extracellular _____ _____ and an increase in extracellular _____ _____.

ANSWER fluid volume . . . solute concentration

QUESTION 100 At osmotic equilibrium, fluid shifts in the dehydrated patient will have occurred which result in *an increase/a decrease* (circle one) in volume in both intracellular and extracellular compartments, together with *an increase/a decrease* (circle one) in solute concentration in both compartments.

ANSWER a decrease . . . an increase

ITEM 28 TREATMENT OF DEHYDRATION

The goal of therapy in the case of the dehydrated patient is to restore total body water to normalcy by fluid replacement, either orally or intravenously. If replacement is intravenous, a hypotonic solution, rather than plain water, is indicated in order to prevent hemolysis of red blood cells. A typical hypotonic solution that is used is 2.5% dextrose in 1/4 normal saline, a solution which becomes increasingly more hypotonic as the dextrose is metabolized.

Intravenous replacement uses a hypotonic solution rather than water because water would _____ red blood cells.	QUESTION 101

hemolyze   ANSWER

In view of the fact that 5% dextrose and 0.9% saline are isotonic with body fluids, it follows that a hypotonic solution would be more *dilute/concentrated* (circle one) than either 5% dextrose or 0.9% saline.	QUESTION 102

dilute   ANSWER

FLUID SHIFTS DUE TO HYPERTONIC INFUSION

ITEM 29

The effect of excessive intravenous infusion of hypertonic NaCl is similar to that of dehydration insofar as tonicity of body fluids is concerned. (Excessive hypertonic infusion and dehydration differ, of course, in that total body water is increased in the former and decreased in the latter.)

Because the solute concentration of extracellular fluid is made up largely of sodium and chloride, intravenous infusion of sodium chloride is commonly employed when replacement therapy is needed. Consider what happens if 2 liters of 4.5% sodium chloride are infused into a patient. Such a solution is drastically hypertonic, and its infusion produces a situation diametrically opposite to that of overhydration. The hypertonic saline solution creates a solute excess in the extracellular compartment. Fluid shifts rapidly from intracellular space to extracellular space as osmotic equilibrium is achieved.

Infusion of 2 liters of 4.5% saline makes plasma hypertonic to all fluids; accordingly, fluid will shift from A. extracellular space to intracellular space B. intracellular space to extracellular space	QUESTION 103

The correct answer is B. The first effect of an infusion of 2 liters of 4.5% saline is an increase in the tonicity of serum. Because serum has become hypertonic to other fluids, water moves from all other compartments into the vascular system in order to dilute the excess solute and restore osmotic equilibrium.	ANSWER

ITEM 30 SEQUENTIAL EFFECTS OF HYPERTONIC SALINE INFUSION ON FLUID VOLUME AND SOLUTE CONCENTRATION

Figure 16 illustrates the relationship between water volume and solute concentration as osmotic equilibrium is being achieved in the patient infused with 2 liters of 4.5% sodium chloride, described in Item 29.

Frame #1 shows the initial condition prior to the saline infusion. Extracellular fluid of 15 liters; intracellular fluid, 25 liters; solute concentration in both compartments, 300 mOsm./L.

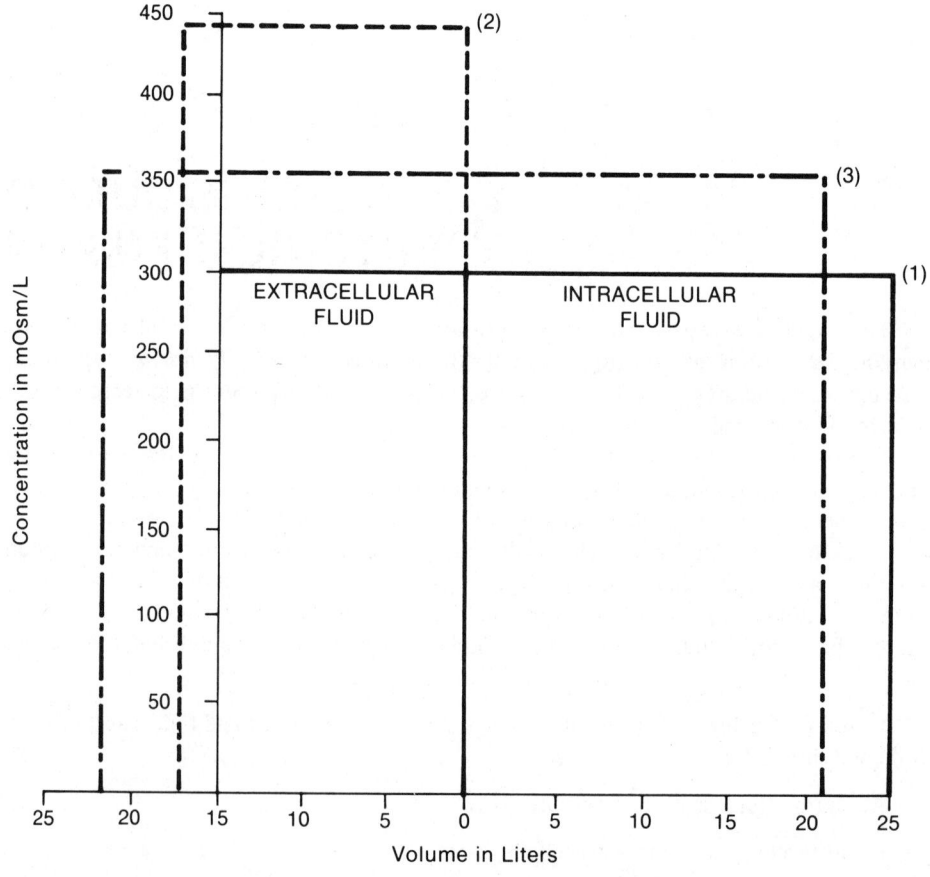

Figure 16.
(1) Solid line (———), frame #1, represents original condition of normal volume and concentration distribution.
(2) Dashed line (------), frame #2, represents the instantaneous effect of infusion of 2 liters of 4.5% sodium chloride.
(3) Dot-dash line (— · —), frame #3, represents final condition after osmotic equilibrium has been achieved.

Frame #2 depicts the instantaneous effect of the infusion with 2 liters of 4.5% saline: expansion of extracellular volume to 17 liters, and a tremendous increase of tonicity to 440 mOsm./L.

At osmotic equilibrium, illustrated in Frame #3, fluid shifts from inside the cells to expand extracellular fluid volume to 21 liters and contract intracellular fluid volume to 21 liters. At equilibrium, the solute concentration of both compartments is equalized at 357 units.

The first effect of infusing a patient with 2 liters of hypertonic saline is an _____ in plasma volume.

QUESTION 104

---

increase

ANSWER

In the case described in Item 29, because fluid shifts from intracellular space to extracellular space, laboratory analyses would show a _____ in all blood values except sodium and chloride.

QUESTION 105

---

decrease

ANSWER

When osmotic equilibrium is reestablished after infusion with hypertonic saline solution, fluid shifts that have occurred cause the volume of extracellular fluids to _____, the volume of intracellular fluids to _____, and the solute concentration of both compartments to _____.

QUESTION 106

---

increase ... decrease ... increase

ANSWER

ADJUSTMENT TO A HYPERTONIC SALINE SOLUTION

ITEM 31

With infusion of a hypertonic solution, serum immediately becomes hypertonic to all other body fluids. The excess salt content of the extracellular fluid inactivates the aldosterone-sodium retention mechanism, and the urinary system can then act to remove the excess sodium.

Before water excretion can stablize, secretion of antidiuretic hormone is adjusted toward restoring the tonicity of body fluids to normal values. Hypertonicity of extracellular fluid effects increased ADH secretion and consequent reabsorption of water from the kidney tubules. While the additional water dilutes extracellular fluid until the kidney restores sodium levels to normal, this extra volume also creates a potentially dangerous hypervolemia. With increased water reabsorption, the kidney elaborates a concentrated, low-volume urine.

QUESTION 107　Increased sodium content of body fluids decreases secretion of the sodium-retaining hormone _____.

ANSWER　aldosterone

QUESTION 108　Hormonal adjustments to hypertonic saline infusion include a shutdown in _____ secretion followed by decreased sodium reabsorption and an increase in _____ _____ secretion, after which there is an increase in water reabsorption.

ANSWER　aldosterone ... antidiuretic hormone

QUESTION 109　Increased fluid volume due to water reabsorption can bring on a hazardous condition of _____.

ANSWER　hypervolemia

ITEM 32　OVERHYDRATION

Water excess may create a problem of *hypo-osmolarity*, or abnormally low concentration of extracellular fluids. In patients with water excess, total body electrolytes are normal, although serum values are low, due to hemodilution. Water intoxication may result from excessive oral or IV intake or from any renal impairment that prevents water excretion. Overhydration can also result from prolonged administration of morphine sulfate, which increases antidiuretic hormone secretion, or from adrenal insufficiency with reduced aldosterone secretion. In addition, hypo-osmolarity of body fluids may develop from replacement of large volumes of sodium lost through vomiting and nasogastric suctioning with plain water instead of water and electrolytes. In any case, in overhydration the osmotic pressure of extracellular fluids is initially low, and fluid shifts to the intracellular space to equalize osmotic pressure until the excess water can be eliminated. Finally, the ADH mechanism responds to the presence of hypotonic ECF by diminishing the secretion of antidiuretic hormone; the kidneys reabsorb less water; glomerular filtration increases; and urinary output increases, leading to polyuria.

QUESTION 110　Hypotonicity of serum brought on by excessive water intake causes fluid to shift from the circulatory system into the _____ space.

ANSWER　intracellular

Hypo-osmolarity may arise from

(1) _____ impairment

(2) excessive intravenous infusion with a _____ solution.

(3) _____ lavage

QUESTION 111

renal ... hypotonic ... gastric

ANSWER

The kidneys respond to overhydration by increasing _____ filtration and _____ output following inhibition of ADH secretion.

QUESTION 112

glomerular ... urinary

ANSWER

SIGNS AND SYMPTOMS OF OVERHYDRATION

ITEM 33

A conspicuous sign of hypo-osmolarity is polyuria, which results from inhibition of release of ADH and consequent diminished renal reabsorption of water. Hypo-osmolarity may also bring on cerebral edema, which is manifested by twitching, hyperirritability, and disorientation. Decreased serum sodium levels (hyponatremia) in the overhydrated patient produce nausea and vomiting. One sign of hyponatremia is a laboratory finding of serum sodium levels below 120 mEq./L., as contrasted with normal levels of 132 to 142 mEq./L.

An overhydrated patient with normal kidney function will have decreased renal _____ and increased urinary output, leading to _____.

QUESTION 113

reabsorption ... polyuria

ANSWER

Disorientation in the overhydrated patient occurs because of _____ edema.

QUESTION 114

cerebral

ANSWER

Gastrointestinal symptoms of hyponatremia are _____ and _____.

QUESTION 115

nausea ... vomiting

ANSWER

QUESTION 116 Whereas normal serum sodium levels range from _____ to _____ mEq./L., laboratory values in cases of hypontremia will fall below _____ mEq./L.

ANSWER 132 ... 142 ... 120

ITEM 34 FLUID VOLUME AND CONCENTRATION IN THE OVERHYDRATED PATIENT

Figure 17 illustrates the relationship between water volume and solute concentration in a hypothetical situation in which a patient has been accidentally infused with two liters of plain water.

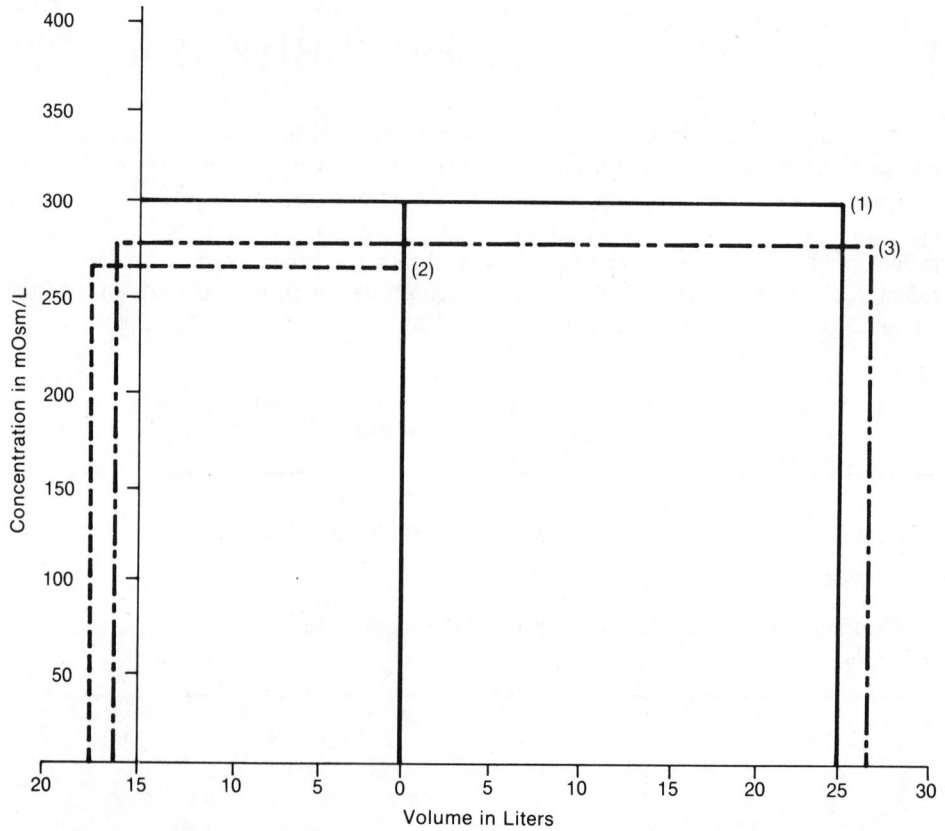

Figure 17.
(1) Solid line (——), frame #1, represents original condition of normal volume and concentration distribution.
(2) Dashed line (------), frame #2, represents the instant effect of excessive fluid intake.
(3) Dot-dash line (— · —), frame #3, represents final condition after osmotic equilibrium has been achieved.

Frame #1 represents the initial normal distribution of fluids and solutes. Water volume in ICF stands at 25 liters; in ECF, at 15 liters. Both compartments are equal in solute concentration: 300 mOsm./L.

Frame #2 depicts the instantaneous effect of the infusion of two liters of water. Serum volume has expanded to a total of 17 liters. Simultaneously, concentration of solutes in extracellular space has been diluted to 265 mOsm./L.

Frame #3 shows the final osmotic equilibrium. Water has shifted into cells until osmotic pressures are equalized in both compartments, with a balanced solute concentration of 285 mOsm./L. Osmosis brings about a final ECF volume of slightly less than 16 liters and an ICF volume of slightly more than 26 liters. Thus, until the kidneys can excrete the overload, the water excess is shared by both compartments.

The normal distribution of water in a human being is _____ liters in intracellular fluid and _____ liters in extracellular fluid; normal solute concentration is equal in both compartments standing at _____ mOsm./L.	QUESTION 117
25 . . . 15 . . . 300	ANSWER
Expansion of serum volume is accompanied by dilution of solute concentration in plasma, making extracellular fluids relatively *hypertonic/hypotonic* (circle one) to intracellular fluids.	QUESTION 118
hypotonic	ANSWER
As osmotic equilibrium is taking place in the overhydrated patient, solute concentration increases in the _____ compartment and decreases in the _____ compartment.	QUESTION 119
extracellular . . . intracellular	ANSWER

TREATMENT OF OVERHYDRATION

ITEM 35

Overhydration does not occur under normal conditions both because normal intake satisfies thirst and because the healthy kidney rapidly excretes any excess water. Under abnormal conditions, however, the kidney may be unable to excrete a water load, and the body may need to be protected against a dangerous level of cerebral edema.

Under clinical circumstances, overhydration is typically treated by infusion of a hypertonic solution to effect an osmotic shift of water from cells vulnerable to overhydration—in this case, cells of the brain. Infusion of hypertonic mannitol may be

78   FLUID BALANCE

employed to produce osmotic diuresis. Mannitol is filtered by the glomerular membrane, but it is neither reabsorbed nor excreted. As glomerular filtrate increases in osmolarity because of the increased osmotic pressure of mannitol, water shifts from peritubular capillaries into glomerular filtrate, thus reducing the hypo-osmolarity of extracellular fluids. The patient excretes a larger and more dilute volume of urine. As extracellular tonicity is restored, intracellular hypotonicity is correspondingly adjusted.

QUESTION 120   Cells which are particularly vulnerable to overhydration are cells of the _____.

ANSWER   brain

QUESTION 121   Infusion of hypertonic mannitol effects _____ diuresis.

ANSWER   osmotic

QUESTION 122   As mannitol raises the osmotic pressure of glomerular filtrate, water shifts from _____ capillaries into glomerular filtrate.

ANSWER   peritubular

QUESTION 123   Healthy kidneys respond to overhydration by putting out

A. small quantities of concentrated urine

B. small quantities of dilute urine

C. large quantities of dilute urine

D. large quantities of concentrated urine

ANSWER   The correct answer is C, large quantities of dilute urine.

ITEM 36   INTAKE AND OUTPUT RECORDS

Intake and output (I and O) records are kept for many hospital patients whose conditions warrant monitoring of fluid status. Some patients may need an increased fluid intake; some, restricted intake; still others may need continual assessment to maintain fluid balance.

For patients with bacterial and viral infections, fluids are encouraged, both to wash out the infectious agent and to replace insensible fluid losses imposed by the rapid respiration and diaphoresis which accompany fever. Restriction of fluid intake, on the other hand, may be indicated in cases of renal shutdown, edema, or congestive heart failure.

Intake and output recording is of special importance for patients vulnerable to _____ _____.

QUESTION 124

fluid imbalance

ANSWER

Febrile patients require adequate fluid intake to replace water lost by increased _____ and _____.

QUESTION 125

respiration . . . diaphoresis

ANSWER

In cases of congestive heart failure, *an increase/a restriction* of fluid intake may be indicated.

QUESTION 126

a restriction

ANSWER

FORCING FLUIDS IN INFECTIONS OF THE URINARY TRACT

ITEM 37

Forcing fluids is particularly important in patients who have infections of the urinary tract; in those immobilized for long periods of time; and in those suffering from certain respiratory disorders. In cases of infections of the urinary tract, increased water intake is obviously important for washing out the infectious agent.

Three of the more common cases in which forcing fluids is indicated are (1) infections of the _____ _____, (2) long periods of _____, and (3) _____ disorders.

QUESTION 127

urinary tract . . . immobilization . . . respiratory

ANSWER

FLUID BALANCE

QUESTION 128 When the urinary tract becomes infected, forcing fluids aids in _____ _____ the infectious agent.

ANSWER washing out

ITEM 38 FORCING FLUIDS IN IMMOBILIZED PATIENTS

Patients immobilized for long periods of time lose calcium from bone, a phenomenon which occurs when normal bone stress and tension are diminished. Increased serum calcium in immobilized patients sometimes tends to crystallize as it is excreted through the renal pelvis. Renal calculi, or kidney stones, form and lodge in the kidney, renal pelvis, or ureter. Dehydration aggravates the tendency to calculi formation, whereas adequate hydration and adequate urinary output may prevent this complication.

QUESTION 129 Long-immobilized patients lose _____ from bone because of decrease of normal bone _____.

ANSWER calcium . . . stress

QUESTION 130 Kidney stones are formed from the _____ of serum _____.

ANSWER crystallization . . . calcium

QUESTION 131 An aggravating factor in the formation of kidney stones is _____, a condition which can be offset by adequate fluid intake.

ANSWER dehydration

FORCING FLUIDS WITH RESPIRATORY DISORDERS

ITEM 39

Ensuring adequate hydration is important in patients with respiratory disorders characterized by increased secretions. For example, patients with emphysema or chronic obstructive pulmonary disease may have difficulty expectorating the excess secretions unless these secretions are sufficiently liquefied by adequate hydration.

Increased fluid intake is an aid to patients with such respiratory disorders as emphysema, in that the fluid _____ respiratory _____ so that they can be expectorated.

QUESTION 132

---

liquefies ... secretions

ANSWER

RESTRICTING FLUIDS

ITEM 40

In contrast to conditions calling *for* fluids, partial or complete renal shutdown requires a restricted fluid intake. When urinary output is greatly reduced or absent, normal intake of fluid expands the extracellular compartment. Such expansion of extracellular fluid in the oliguric or anuric patient carries with it the risk of hypervolemia, increased blood pressure, and a possible circulatory overload.

Patients with congestive heart failure also may need to be protected against water overload and hypervolemia, for these conditions impose an additional workload on the heart.

Two of the more common pathological conditions requiring restriction of fluid intake are _____ _____ and _____ _____ _____ .

QUESTION 133

---

renal shutdown ... congestive heart failure

ANSWER

The adverse effect of fluid intake on extremely oliguric or anuric patients is expansion of the _____ compartment, resulting in _____ .

QUESTION 134

---

extracellular ... hypervolemia

ANSWER

One danger normal fluid intake may pose for the patient with congestive heart failure is an excessive _____ on the heart brought on by _____ .

QUESTION 135

---

workload ... hypervolemia

ANSWER

ITEM 41 INTAKE AND OUTPUT RECORDS FOR POSTSURGICAL PATIENTS

I and O records are frequently kept for a day or two for postsurgical patients. Immediately following surgery, the patient may be somewhat dehydrated, for several reasons. Some fluid deficit may be attributed to the requirement that the patient receive nothing by mouth (written *NPO*) prior to surgery; some results from fluid losses sustained during the surgical procedure. Another cause of a degree of postoperative dehydration is the use of such parasympatholytic drugs as atropine and scopolamine, which are employed during surgery to reduce respiratory and salivary secretions. Use of parasympatholytic drugs calls for monitoring of intake and output, beause they produce as side effects a temporary reduction in gastrointestinal activity and atony of the urinary bladder. If the patient is unable to take fluids by mouth until gastrointestinal activity returns, intravenous replacement is utilized. Until bladder function returns, the patient may have an indwelling (Foley) catheter to prevent urinary stasis.

QUESTION 136 Postsurgical patients tend to be dehydrated because of the _____ restriction prior to surgery and _____ during surgery.

ANSWER NPO (nothing by mouth) . . . fluid losses

QUESTION 137 As consciousness returns, the postsurgical patient feels thirsty because of side effects of _____ drugs.

ANSWER parasympatholytic

QUESTION 138 Two side effects of atropine and scopolamine that call for I and O records postoperatively are decreased _____ activity and diminished _____ output.

ANSWER gastrointestinal . . . urinary

QUESTION 139 To prevent fluid imbalance, the postsurgical patient may receive _____ fluids and have an indwelling _____ for output.

ANSWER intravenous . . . catheter

INTAKE AND OUTPUT SHEETS

ITEM 42

An intake and output sheet provides an accurate means of assessing an individual patient's fluid status. Intake, recorded in milliliters (ml.) or cubic centimeters (cc.), includes fluid derived from any source: by mouth (per os, or PO), intravenous infusion (IV), blood transfusions, and formula feedings.

Output, as recorded on the I and O sheet, includes urinary output, gastric suction, vomitus, watery stools, and excessive drainage from a wound or surgical site. I and O totals are recorded on the patient's chart periodically.

Figure 18 shows a typical form commonly used as an I and O sheet. Note that provision is made for recording at the end of each shift and at the end of 24 hours.

INTAKE	2pm	10pm	6am	24hr	2pm	10pm	6am	24hr
P.O. Fluids								
I.V. Fluids								
Blood								
Formula								
Total								
OUTPUT	2pm	10pm	6am	24hr	2pm	10pm	6am	24hr
Urine								
Gastric Drainage								
Stools								
Total								

Figure 18. An intake-and-output sheet, with columns for entries at specified times.

Intake recording should include fluids taken by _____ or _____, as well as _____ transfusions and _____ feedings.

QUESTION 140

---

mouth (PO, per os) ... intravenously (IV) ... blood ... formula

ANSWER

Output records include fluid loss in _____ output, gastric suction, _____, watery _____, and wound _____.

QUESTION 141

---

urinary ... vomitus ... stool (diarrhea) ... drainage

ANSWER

ISOTONIC VOLUME EXPANDERS

ITEM 43

When intravenous (IV) fluids are indicated, the choice of fluids is determined by the patient's specific need for a hypotonic, isotonic, or hypertonic solution and by the needs for various replacement electrolytes. In some instances—in the case of hemorrhage, for example—an isotonic volume expander may be indicated. In such a situation, whole

84 FLUID BALANCE

blood, dextran, or 5% dextrose in normal saline ($D_5 N/S$) may be used. Dextrose is added to isotonic saline for its energy value. If $D_5 N/S$ is infused slowly, dextrose adds to the osmolar value only temporarily. As dextrose goes into cells and is metabolized, the molecular particles are removed from extracellular fluids. Ultimately, the dextrose solution contributes only water and carbon dioxide as final metabolites. Thus $D_5 N/S$, hypertonic with body fluids at the time of infusion, becomes isotonic as carbon dioxide is removed and water retained.

QUESTION 142 $D_5 N/S$ is the standard abbreviation for the frequently used intravenous volume expander _____ in _____ saline.

ANSWER 5% dextrose . . . normal (isotonic)

QUESTION 143 If dextrose is added to normal saline, the solution is initially hypertonic, but it becomes isotonic as the dextrose molecule metabolizes to _____ and _____.

ANSWER carbon dioxide . . . water

QUESTION 144 A typical clinical situation in which $D_5 N/S$ is indicated is the case of a patient with _____.

ANSWER hemorrhage

QUESTION 145 The choice of solution used for intravenous fluid replacement depends upon a given patient's need for a _____, _____, or _____ volume expander.

ANSWER hypotonic . . . hypertonic . . . isotonic (in any order)

> TRACK TWO. Items 44 and 45 deal with calculation of isotonic solutions. Students in accelerated programs doing Track One may omit the next two Items and go directly to Item 46.

CALCULATION OF ISOTONIC SOLUTIONS: NORMAL SALINE

ITEM 44

The proportion of solute to solvent which makes up an intravenous solution is, in actual practice, calculated by the drug supply houses. For those interested in understanding how the particular proportions are arrived at, this and the next Item will analyze, as illustrations, the calculations that establish that both 0.9% saline and 5% dextrose are isotonic solutions.

Isotonic solutions are calculated on the basis of the molecular weight of the solute. The molecular weight of sodium chloride is 58. One mol (58 grams) of NaCl in 1 liter of water constitutes a 2-osmolar solution, because sodium chloride forms 2 ions and each ion contributes one osmolar unit. Fifty-eight *milli*grams of NaCl in 1 liter of water is a 2-milliosmolar solution (2 mOsm./L.). Since body fluids are a 300 milliosmolar solution (300 mOsm./L.), it would take 150 times the 58 milligrams of NaCl (300 divided by the 2 milliosmols of NaCl) to have a solution isotonic with body fluid. Therefore, 8700 mg. (150 X 58) NaCl per liter would be isotonic with body fluids. Converting to grams, 8700 mg./L. = 8.7 gm./L. = 0.87 gm./100 ml. For simplification, 0.9 gm. is generally used instead of 0.87. Thus, 0.9 grams of sodium chloride per 100 milliliters (0.9 gm.%) is normal, or isotonic, saline.

The solute concentration of an isotonic solution is determined by _____ weight.

QUESTION 146

---

molecular

ANSWER

Fifty-eight milligrams of sodium chloride in a liter of solution contributes 2 milliosmolar units of solute because sodium chloride forms _____ in solution.

QUESTION 147

---

two ions

ANSWER

Since body fluids are (approximately) a 300-milliosmolar solution and 38 mg. sodium chloride in 1 liter of water is a 2-milliosmolar solution, it is necessary to multiply the molecular wieght of sodium chloride by _____ to determine the amount of NaCl in milligrams which would be isotonic with body fluids.

QUESTION 148

150

ANSWER

The amount of sodium chloride per liter needed to form a solution which is isotonic with body fluids is 58 (molecular weight of NaCl) X 150 = 8700 mg.; 8700 mg./L. = _____ gm./L., or _____ gm./100 ml., or, for simplification, _____ gm.%.

QUESTION 149

8.7 ... 0.87 ... 0.9

ANSWER

ITEM 45 — CALCULATION OF ISOTONIC SOLUTIONS: 5% DEXTROSE

Using the same system of calculation as that used in the case of 0.9% saline, one can see why 5% dextrose is isotonic. The molecular weight of dextrose is 180. Since dextrose does not dissociate, 180 grams of dextrose contribute one osmolar unit of solute. The amount of dextrose needed to form a solution which is isotonic with body fluid is the molecular weight of dextrose, 180, times the 300 milliosmols in body fluids: $180 \times 300 = 54{,}000$ mg. Converting milligrams to grams, 54,000 mg./L. = 54 gm./L., or 5.4 grams per 100 milliliters. Thus, for purposes of simplification, a 5% solution of dextrose is isotonic.

QUESTION 150 — One hundred and eighty grams of dextrose constitute a *one/two* (circle one) osmolar solution because dextrose *does/does not* (circle one) dissociate.

ANSWER — one ... does not

QUESTION 151 — Multiplying a milligram molecular weight of dextrose by 300 gives a quantity of dextrose which approximates the tonicity of _____.

ANSWER — body fluids

QUESTION 152 — The amount of dextrose per liter needed to form a solution which is isotonic with body fluids is 180 (molecular weight of dextrose) $\times$ 300 = 54,000 mg.; 54,000 mg./L. = _____ gm./L., or _____ gm./100 ml., or, for simplification, _____ % dextrose.

ANSWER — 54 ... 5.4 ... 5

> Students taking the accelerated Track One proceed with Item 46.

ITEM 46 — BALANCED SALT SOLUTIONS

Intravenous infusions designed to remedy electrolyte imbalances are termed *balanced solutions*. Balanced solutions contain electrolyte concentrations which approximate those of body fluids and include sodium, potassium, calcium, chloride, bicarbonate, and phosphate ions. Included in this group of solutions is the well-known Ringer's lactate solution, which is frequently used as a postoperative intravenous replacement. Balanced solutions may also be preferred if the infusion is to be continued over several days. Ringer's lactate solution is commonly ordered following abdominal surgery, when food and fluid by mouth are withheld until peristalsis returns.

Intravenous solutions which contain electrolyte concentrations approximating the concentrations of body fluids are called _____ solutions.	QUESTION 153
balanced	ANSWER
Ringer's lactate solution, used as a postoperative intravenous replacement, is particularly effective after _____ surgery, when oral intake is withheld.	QUESTION 154
abdominal	ANSWER
In addition to sodium and chloride, ions contained in balanced solutions include _____, _____, _____ and _____.	QUESTION 155
potassium . . . calcium . . . bicarbonate . . . phosphate (in any order)	ANSWER

FLOW RATE OF INTRAVENOUS INFUSIONS

ITEM 47

Volume of flow of an intravenous infusion is prescribed by the physician. A typical example is IV D_5 N/S (5% dextrose in normal saline), 100 ml./hr. In a 24-hour period, the patient on such a flow rate would receive 2400 milliliters.

To calculate the *drip rate*, it is necessary to know the drip rate in drops per milliliter (abbreviated *gtts/ml.*) of the particular tubing used. Baxter (Travenol) IV tubing delivers 10 gtts/ml.; Abbott, 15 gtts/ml.; Cutter, 20 gtts/ml. Drip rate is calculated according to the following formula:

$$\text{gtts/minute} = \frac{\text{total volume in ml.} \times \text{drip rate of tubing/ml.}}{\text{number of hours} \times \text{minutes per hour}}$$

Substituting the physician's order mentioned in the first paragraph and the drip rate of Baxter tubing in the formula, one can calculate the number of drops per minute as follows:

$$\text{gtts/minute} = \frac{100 \text{ ml.} \times 10 \text{ (Baxter tubing)}}{1 \text{ (hour)} \times 60 \text{ (minutes per hour)}} = \frac{1000}{60} = 16\frac{2}{3}$$

The drip rate is 16 2/3 drops per minute, or, for simplification, 17 drops per minute.

88 FLUID BALANCE

Consider a second example. Suppose the physician ordered 1000 ml. of Ringer's lactate solution (abbreviated RL), to be infused over three successive eight-hour periods. The order might read: "IV 1000 ml. RL q 8 hrs. X 3." Given the physician's order and using Baxter tubing in the formula, one calculates the drip rate as:

$$\text{gtts/minute} = \frac{1000 \times 10}{8 \times 60} = \frac{10{,}000}{480} = 21 \text{ drops per minute}$$

QUESTION 156 The formula for calculating flow rate for intravenous infusion in drops per minute is as follows:

$$\text{gtts/minute} = \underline{\hspace{3cm}}$$

ANSWER

$$\frac{\text{total volume in ml.} \times \text{drip rate of tubing/ml.}}{\text{number of hours} \times 60 \text{ (min./hr.)}}$$

QUESTION 157 Calculate the flow rate for an order for IV 1000 ml. Ringer's lactate solution every 8 hours X 3, using Cutter tubing. (The drip rate of Cutter tubing can be found in this Item.)

ANSWER

$$\text{gtts/minute} = \frac{1000 \text{ ml.} \times 20 \text{ gtts/ml.}}{8 \text{ hrs.} \times 60 \text{ min.}} = \frac{20{,}000}{480} = 42 \text{ gtts/min.}$$

QUESTION 158 Calculate the flow rate for an order for IV 1000 ml. D_5 ½ N/S at 75 ml./hr. X 3, using Baxter tubing. (The drip rate of Baxter tubing can be found in this Item.)

ANSWER

$$\text{gtts/minute} = \frac{75 \text{ ml.} \times 10 \text{ gtts/ml.}}{1 \text{ hr.} \times 60 \text{ min}} = \frac{750}{60} = 12.5 \text{ gtts/min.}$$

ITEM 48 TIME STRIPS FOR INTRAVENOUS INFUSIONS

Intravenous bottles are frequently "time-stripped" with an appropriate marker for determining at a glance whether the infusion is flowing on schedule. Figure 19 illustrates a typical time strip to be attached to the infusion bag. Milliliters, in increments of 50, are shown on the right. Starting time is indicated at the top, and successive hours are marked at various increments on the 1000-milliliter strip. Appropriate bars are marked for infusions to run 4, 6, 8, or 12 hours. For example, a 1000-ml. infusion, hung at 8:00 A.M. and to run 8 hours, will have infused 125 ml. at 9:00 A.M.; 250 ml. at

10:00 A.M.; and 500 ml. at noon. If the infusion is running too fast or too slowly, appropriate adjustments can be made in the flow rate.

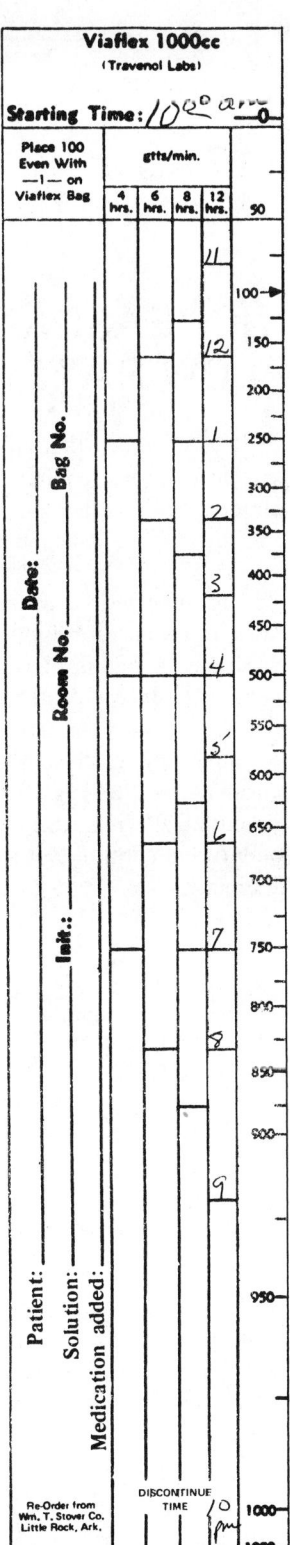

Figure 19. Time strip for intravenous infusion. The strip is attached to an infusion bag and the rate of emptying is measured on the milliliter scale. (Courtesy of Travenol Labs, Deerfield, Illinois.)

FLUID BALANCE

QUESTION 159 Using a time strip, calculate the following: If a 1000-ml. solution is hung at 6:00 A.M. to run for 6 hours, approximately 165 to 170 ml. of solution will have been infused at _____ A.M.; 500 ml., at _____ A.M.; and 860 ml., at _____ A.M.

ANSWER 7:00 ... 9:00 ... 11:00

ITEM 49 INCREASING INTRAVENOUS FLOW RATE

It may happen that an intravenous infusion is flowing behind schedule, and that the physician's order requires that the flow rate not exceed a certain number of drops per minute if behind schedule. A physician might impose a restriction on increase of flow rate if a potent drug is added to the infusion or if there is some possibility of circulatory overload.

The following hypothetical case illustrates how much to increase the flow rate of an infusion which is behind schedule and remain within a restriction imposed by the physician. In this illustration the order reads: "IV 1000 ml. RL q 8 hrs., with 40 mEq. KCl, not to exceed 75 gtts/min." Suppose that the infusion started at 10:00 P.M. and that the patient slept on the tubing (in this case Baxter tubing), causing a kink which prevented the calculated flow rate from infusing. At 2:00 A.M., instead of 500 ml. remaining in the bag, 800 ml. remains to be infused, in 4 hours. The formula for computing gtts/min. can be applied to gauge the drip rate for the remaining time so that the infusion continues at a faster rate, but one that will not exceed the restriction of 75 gtts/min.:

$$\text{gtts/min.} = \frac{800 \text{ ml. (yet to be infused)} \times 10 \text{ (Baxter tubing)}}{4 \text{ (hours remaining)} \times 60 \text{ (minutes per hour)}}$$

$$= \frac{8000}{240} = 33 \text{ gtts/min.}$$

The accelerated flow rate of 33 drops per minute is well below the restriction of 75 drops per minute.

QUESTION 160 An order reads: "IV 1000 ml. D_5 N/S, with 40 mEq. KCl, at 100 ml./min., not to exceed 50 gtts/min. if running behind." The IV was hung at 12:00 midnight. At 5:00 A.M., 200 ml. had been infused. Calculate the adjusted flow rate (Baxter tubing).

ANSWER $\text{gtts/min.} = \frac{800 \text{ ml} \times 10}{5 \text{ hrs.} \times 60} = \frac{8000}{300} = 27$ (well below restriction of 50 gtts/min.)

CAPILLARY DYNAMICS ITEM 50

Previous Items in this Unit dealt with fluid shifts that occur between compartments in order to maintain osmotic equilibrium in response to varying intake and output. The remaining Items will deal with fluid shifts across the capillary membrane in both normal and pathological states.

Exchange of fluid between plasma and interstitial fluid allows an inflow of nutrients to the cells. On the other hand, fluid shifts from interstitial space into the capillary permit removal of metabolic products from the cellular environment. Exchange of fluid between capillaries and interstitial space is referred to collectively as *capillary dynamics*. Capillary dynamics embraces the sum total of all the forces which tend to move fluid in one direction as balanced against the forces which tend to oppose and redirect flow in the opposite direction. The major means of fluid transfer are diffusion and osmosis.

Capillary dynamics accounts for outflow of fluid from serum to _____ space as well as inflow of fluid from _____ space to serum.	QUESTION 161
interstitial . . . interstitial	ANSWER
The mechanisms of membrane transport which capillary dynamics utilizes are _____ and _____.	QUESTION 162
diffusion . . . osmosis	ANSWER

PRESSURE FACTORS INVOLVED IN CAPILLARY DYNAMICS ITEM 51

Figure 20 illustrates the forces at the capillary membrane that tend to move fluid either outward or inward through the membrane.

Three factors tend to move fluid through pores in the capillary wall from serum to interstitial fluid: (1) capillary pressure, (2) tissue pressure (i.e., negative interstitial fluid pressure), and (3) interstitial fluid colloid osmotic pressure.

A single factor moves fluid from interstitial fluid to serum—namely, serum colloid osmotic pressure.

92 FLUID BALANCE

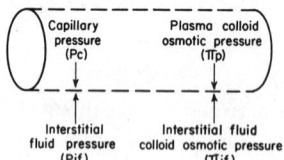

Figure 20. Forces operative at the capillary membrane tending to move fluid either outward or inward through the membrane. (From Guyton: *Textbook of Medical Physiology.* 5th Ed. Philadelphia: W. B. Saunders Company, 1976.)

QUESTION 163
Fluid shifts from the capillary into interstitial space because of (1) _____ pressure, (2) _____ pressure, and (3) interstitial fluid colloid _____ pressure.

ANSWER
capillary ... tissue ... osmotic

QUESTION 164
Fluid moves from interstitial space into capillaries as a result of serum _____ _____ pressure.

ANSWER
colloid osmotic

ITEM 52 CAPILLARY PRESSURE

Capillary pressure is the hydrostatic pressure remaining from the pumping action of the heart. Capillary pressure varies with fluctuations in mean arterial blood pressure. Under normal circumstances, the pressure remaining at the arteriole end of the capillary is about 25 mm. Hg. Pressure at the venule end of the capillary is about 10 mm. Hg. The mean capillary pressure is about 17 mm. Hg. The pressure differential in the capillary causes fluid to flow from the arterioles to the venules. When mean arterial blood pressure increases, capillary pressure increases; when mean arterial pressure is reduced, capillary pressure is correspondingly reduced.

QUESTION 165
Blood pressure which remains at the capillary level is termed _____ pressure.

ANSWER
capillary

QUESTION 166
Capillary pressure is higher at the _____ end of the capillary than it is at the _____ end.

ANSWER
arteriole ... venule

Blood flow is unidirectional from arterioles to venules because of pressure _____.	QUESTION 167
differences (gradients)	ANSWER

Capillary pressure level varies with increase and decrease in _____ _____ pressure.	QUESTION 168
mean arterial	ANSWER

INTERSTITIAL FLUID PRESSURE — ITEM 53

Interstitial fluid pressure is normally about −7 mm. Hg. This value is negative because of a certain amount of functional vacuum created as lymph flows from interstitial space into the lymphatic system. Lymph flow is maintained by the action of tissue pressing against lymphatic vessels. This fact, plus the one-way structure of valves, ensures, under normal circumstances, a continual drainage of interstitial fluid as fast as it accumulates.

Interstitial fluid pressure is at a negative value primarily because of _____ _____.	QUESTION 169
lymphatic drainage	ANSWER

SERUM AND INTERSTITIAL FLUID COLLOID OSMOTIC PRESSURES — ITEM 54

Serum colloid osmotic pressure, about 28 mm. Hg, results from the large quantities of non-diffusible plasma protein, chiefly albumin. A small amount of protein which leaks from serum into interstitial space also creates an interstitial fluid colloid osmotic pressure, normally about 4.5 mm. Hg. Both serum colloid osmotic pressure and interstitial fluid colloid osmotic pressure exert a drawing effect on water. Serum proteins, however, exert a much greater osmotic effect than that exerted by the smaller amount of protein present in interstitial fluid, as evidenced by the pressure differences.

QUESTION 170   Serum colloid osmotic pressure results from large quantities of _____ in blood.

ANSWER   albumin

QUESTION 171   Serum colloid osmotic pressure tends to *promote/restrain* (circle one) outward flow of water from capillary beds.

ANSWER   restrain

QUESTION 172   Interstitial fluid colloid osmotic pressure tends to promote _____ flow from the _____ to _____ space.

ANSWER   outward ... capillary ... interstitial

ITEM 55 CAPILLARY FLUID EXCHANGE—SUM OF FORCES

Figure 21 shows the sum of outward and inward forces which determine the direction of fluid shift across the capillary membrane at both the arteriole and venule ends of the capillary. The single inward force, serum colloid osmotic pressure, is the same at both the arteriole and venule ends of the capillary: 28.0 mm. Hg. Two of the three outward forces, interstitial fluid pressure and interstitial fluid colloid osmotic pressure, are likewise the same at both ends of the capillary: 7.0 mm. Hg and 4.5 mm. Hg, respectively. However, capillary pressure, the third outward force, is much higher at the arteriole end of the capillary—25.0 mm.Hg—than it is at the venule end, where it is 9.0 mm. Hg.

The pressures exerted by the several components of total outward force at the arteriole end of the capillary are as follows:

(1) capillary pressure .... 25.0 mm. Hg
(2) interstitial fluid pressure .... 7.0 mm. Hg
(3) interstitial fluid colloid osmotic pressure .... 4.5 mm. Hg
Total outward force at the arteriole end .... 36.5 mm. Hg

At the arteriole end of the capillary, the total outward force of 36.5 mm. Hg exceeds the total inward force of serum colloid osmotic pressure of 28.0 mm. Hg, for a net outward

CAPILLARY FLUID EXCHANGE—SUM OF FORCES

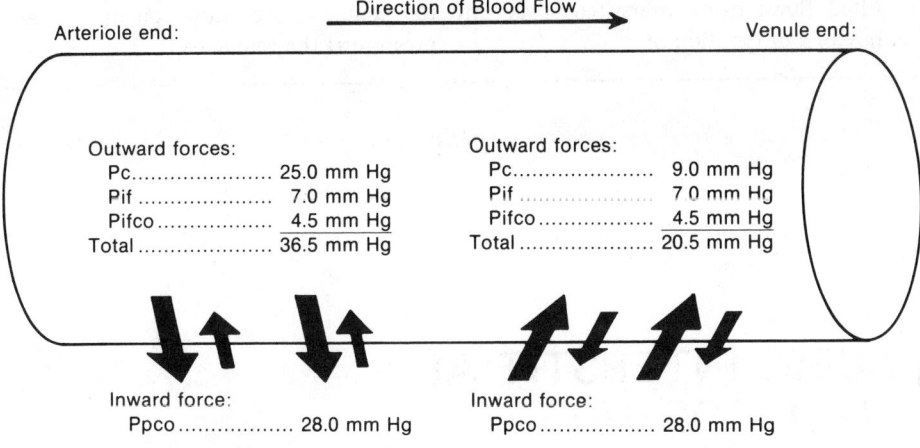

Figure 21. Capillary membrane.
Key:
P_c = capillary pressure
P_{if} = interstitial fluid pressure
P_{ifco} = interstitial fluid colloid osmotic pressure
P_{pco} = plasma colloid osmotic pressure

force of 8.5 mm. Hg (36.5 − 28.0 = 8.5). Therefore, the net fluid shift at the arteriole end is outward from the capillary to interstitial space.

The pressures exerted by the several components of total outward force at the venule end of the capillary are as follows:

(1) capillary pressure	9.0 mm. Hg
(2) interstitial fluid pressure	7.0 mm. Hg
(3) interstitial fluid colloid osmotic pressure	4.5 mm. Hg
Total outward force at the venule end	20.5 mm. Hg

At the venule end of the capillary, the total inward force of serum colloid osmotic pressure, 28.0 mm. Hg, exceeds the total outward force of 20.5 mm. Hg, for a net inward force of 7.5 mm. Hg (28.0 − 20.5 = 7.5). Therefore, the net fluid shift at the venule end is inward from interstitial space to capillary.

QUESTION 173

When the sum of outward pressure components exceeds the force of serum colloid osmotic pressure, fluid flows by osmosis from

A. serum to interstitial fluid

B. interstitial fluid to plasma

ANSWER

The correct answer is A. Serum colloid osmotic pressure is the single inward force moving fluid across the capillary membrane; when this inward force is exceeded by outward forces, fluid moves out of the capillary into interstitial space.

QUESTION 174

At the arteriole end of the capillary, net *outward/inward* (circle one) force is greater, and consequently fluid shifts by osmosis into *serum/interstitial space* (circle one).

ANSWER

outward . . . interstitial space

96 FLUID BALANCE

QUESTION 175 Fluid flows from interstitial space into the venule end of the capillary primarily because of a diminution of _____ pressure at the venule end.

ANSWER capillary

ITEM 56 EXCESS INTERSTITIAL FLUID—EDEMA

One of the most common problems in fluid and electrolyte balance is *edema*. Edema results from abnormal compartmentalization of body fluids because of excess fluid in interstitial space. Actually, edema is present whenever interstitial fluid pressure increases to a positive value.

Edema is associated with many diverse disturbances such as congestive heart failure, extensive burns, allergic reactions, cancer, and dietary protein deficiencies. The syndrome common to all these diverse conditions is an excess of interstitial fluid in the tissue spaces. Edema is essentially a problem in exchange of fluid between the blood and interstitial space. The function of interstitial fluid is the transfer of nutrients from the blood to the cells, and of metabolites from the cells to the blood. These substances transfer rapidly when the quantity of interstitial fluid is small, and slowly when the quantity is too great.

Remember that edema occurs when the normally negative interstitial fluid pressure becomes positive. At that point, excess interstitial fluid accumulates in tissue spaces, bringing on such complications as gangrenous limbs, compressed blood vessels, depressed respiratory exchange, and a variety of other states.

QUESTION 176 Edema is due to the accumulation of fluid in _____ _____.

ANSWER interstitial spaces

QUESTION 177 Edema occurs when interstitial fluid pressure loses its _____ value.

ANSWER negative

QUESTION 178 The characteristic manifestation of edema is a _____ of tissue.

ANSWER swelling

Among the more common pathological conditions associated with edema are (1) _____ _____ _____ , (2) _____ , and (3) _____ .

QUESTION 179

congestive heart failure . . . cancer . . . burns
(also: allergic reactions and protein deficiencies)

ANSWER

Nutrients and metabolites transfer rapidly when the quantity of interstitial fluid is _____ , but slowly when the quantity of interstitial fluid is _____ .

QUESTION 180

small . . . large

ANSWER

FACTORS WHICH NORMALLY PREVENT EDEMA FORMATION

ITEM 57

When interstitial fluid pressure changes from negative to positive, edema develops. Two factors ordinarily keep interstitial fluid pressure negative: (1) reabsorption of fluid into the circulatory system and (2) diffusion of fluid and protein from tissue spaces into lymphatic capillaries.

The combined forces of normal capillary dynamics maintain an average negativity of −7 mm. Hg in interstitial fluid pressure. This constitutes a 7 mm. Hg safety element.

Tissue pressure remains negative because of _____ of fluid into capillaries and _____ of fluid into lymphatics.

QUESTION 181

reabsorption . . . diffusion

ANSWER

The fact that tissue fluid pressure is normally around −7 mm. Hg provides a _____ _____ against edema formation.

QUESTION 182

safety factor

ANSWER

ITEM 58 — LYMPHATIC DRAINAGE

Lymphatic vessels provide an accessory route through which interstitial fluids can flow from tissue space into venous blood. Whenever interstitial fluid accumulates in tissue space, the amount and rate of lymph flow increases, thereby reducing the danger of edema formation.

Lymph flow also helps preclude edema by washing out the albumin that constantly leaks in small quantities from serum into the tissue spaces. If allowed to accumulate, albumin will increase the tissue colloid osmotic pressure and bring on edema. The washout process prevents the accumulation of serum protein and thereby removes the threat of abnormally high tissue colloid osmotic pressure.

QUESTION 183 As the volume of interstitial fluid increases, the flow of lymph _____.

ANSWER increases

QUESTION 184 Lymph flow removes excess _____ from tissue space, and also removes serum _____, which leaks through capillaries into tissue space.

ANSWER fluid ... albumin (or protein)

QUESTION 185 The lymphatic system lessens the possibility of edema by preventing a rise in tissue _____ _____ pressure.

ANSWER colloid osmotic

ITEM 59 — HIGH CAPILLARY PRESSURE AS A CAUSE OF EDEMA

Edema can ensue from high capillary pressure, low plasma protein concentration, lymphatic blockage, or increased capillary porosity. The first of these causes, high capillary pressure, can result from any condition involving decreased venous return, such as occurs in congestive heart failure. In congestive heart failure, the inability of the heart to pump blood adequately through the heart leads to venous pooling. As central venous pressure rises, capillary pressure rises, and abnormally large quantities of fluid diffuse from capillaries to interstitial space, constituting a condition of edema.

QUESTION 186

Edema is caused by various disturbances in capillary dynamics, among the more common of which are:

(1) high _____ pressure

(2) low blood _____

(3) _____ blockage

(4) increased _____ porosity

ANSWER

(1) capillary

(2) protein

(3) lymphatic

(4) capillary

QUESTION 187

One of the more common conditions which raises capillary pressure to a point of producing edema is _____.

ANSWER

congestive heart failure

LOW BLOOD PROTEIN AS A CAUSE OF EDEMA

ITEM 60

Under normal conditions, osmotic pressure of serum is higher than that of interstitial fluid, primarily because of the presence of albumin in the blood serum. A decrease in serum proteins, or *hypoproteinemia*, reduces the concentration of total protein and, therefore, the level of serum colloid osmotic pressure. When this happens, the total inward force due to serum proteins is reduced, and fluid tends to remain in interstitial space.

Albumin can be lost from the body in several ways. In extensive burns, fluid that contains protein leaks from the denuded areas of the skin. In nephrosis, albumin filters through the glomerular membrane of the kidney and is lost in urine. A metabolic abnormality called *agammaglobulinemia* can also cause low blood protein, owing to decreased formation of serum globulins. Finally, a simple deficiency in protein intake in the diet can bring on hypoproteinemia. In all of these conditions, the total inward force due to serum proteins is reduced, creating a condition which favors edema formation.

QUESTION 188

The plasma protein most responsible for plasma colloid osmotic pressure is _____.

ANSWER

albumin

| QUESTION 189 | Decreased serum protein promotes edema because of a reduction of serum _____ pressure. |

| ANSWER | colloid osmotic |

| QUESTION 190 | Low blood protein is conducive to edema because decreased serum albumin reduces the pressure of the single force that promotes movement of fluid

A. outward from capillaries to interstitial space
B. inward from interstitial space to serum |

| ANSWER | The correct answer is B. Serum colloid osmotic pressure due to blood protein promotes movement of fluid from interstitial space into the venule end of the capillary. With lowered blood protein, serum colloid osmotic pressure is reduced, and fluid remains in interstitial space, where it forms edema. |

| QUESTION 191 | Besides metabolic abnormalities and dietary insufficiency, other clinical conditions which reduce blood protein and lead to edema include _____ and _____. |

| ANSWER | burns ... nephrosis (in either order) |

ITEM 61 LYMPHATIC BLOCKAGE AS A CAUSE OF EDEMA

Protein substances which leak out of the capillaires into interstitial fluid are normally removed and returned to venous blood through the lymphatic system. Pathological conditions, however, can block the lymphatic vessels. When the lymphatic channel of removal is cut off, albumin collects in tissue spaces, increasing the outward force of negative interstitial pressure and thereby forcing fluid from capillaries into interstitial space.

One common instance of lymphatic blockage occurs when pectoral and axillary lymph nodes are removed, as in a radical mastectomy. Removal of a cancerous breast and adjacent lymph nodes is frequently accompanied by edema in the arm on the side of the excision. A patient who has undergone a radical mastectomy must periodically elevate the affected arm to facilitate venous return and thereby decrease lymphedema.

| QUESTION 192 | Lymphatic channels normally return leaked serum protein to the venous _____. |

| ANSWER | blood (or circulation) |

Lymphatic blockage prevents removal of protein from _____ spaces.	QUESTION 193
interstitial	ANSWER
Accumulations of protein in tissue spaces increases tissue colloid osmotic pressure and, in effect, retards _____ flow of fluid.	QUESTION 194
inward	
Lymphatic blockage and lymphedema commonly follow removal of lymph nodes adjacent to the _____, as in a _____.	QUESTION 195
breast ... radical mastectomy	ANSWER

INCREASED CAPILLARY POROSITY AS A CAUSE OF EDEMA

ITEM 62

Increased capillary porosity is also a cause of edema. Allergy produces capillary dilatation through the release of histamine. As capillaries dilate locally, both serum protein and fluid are lost through the enlarged capillary pores. The loss of fluid and protein produces a congestion of fluid in interstitial spaces, or localized edema.

Increased capillary porosity promotes edema formation because of loss of _____ and serum _____ through the enlarged pores.	QUESTION 196
fluid ... protein	ANSWER
Capillaries dilate in local allergic reactions through the release of _____.	QUESTION 197
histamine	ANSWER

SUMMARY OF UNIT TWO

Various membranes separate total body water into compartments. Fluid within cells is called intracellular fluid; that not within cells is called extracellular fluid. A continual shift of water occurs between these compartments, just as water within the body as a whole is continually interchanged. Both fluids and dissolved particles are in a state of dynamic equilibrium; that is to say, although composition remains relatively stable, there is a continuous mixing of fluids, losses and replacement, resulting in an uninterrupted turnover of molecules and ions.

Interchange of fluid between compartments is due to pressure gradients and to concentration gradients. Hydrostatic pressure gradients result from the pumping action of the heart, blockage of vessels, and so forth. Concentration gradients effect shifts according to the principle of osmosis.

Direction of flow in osmosis is determined by the total body solute concentration in each compartment. Water moves across the semipermeable membranes from an area of high concentration of water (low solute concentration) into an area of low concentration of water (high solute concentration).

When a concentration of particles has the same osmotic pressure as body cells, the solution is called isotonic. A hypotonic solution is one in which the concentration of particles is less than that in cells. A hypertonic solution is one in which the concentration of particles is greater than that in cells. Cells, such as red blood cells, can be suspended in an isotonic solution with no net shift of water either into or out of the cells because there is no difference in the osmotic pressures of the two fluid compartments. When blood cells are suspended in a hypotonic solution, water moves from the solution outside the cell into the intracellular fluid, that is, into the area with the higher osmotic pressure. Conversely, when blood cells are suspended in a hypertonic solution, water moves from the intracellular compartment into the serum.

Total body water is determined by a balance between intake and output. Daily intake from fluids that are ingested and from the metabolism of food usually approximates about 2600 ml. Daily output of water averages about 1500 ml. in urine, 100 ml. lost in feces, and 1000 ml. in insensible water loss through the skin and lungs. When fluid intake is excessive, the kidneys remove this excess by forming a larger quantity of dilute urine. In the case of diminished fluid intake, the kidneys conserve body water by forming a lesser quantity of concentrated urine.

Each kidney is made up of millions of nephrons, each with a similar structure and function. A single nephron is composed of a glomerulus, a tubule, and a collecting duct. The glomerulus is a capsule-like structure enclosing a tuft of capillaries, from which water and electrolytes are filtered to form glomerular filtrate. As glomerular filtrate courses through the tubule, some substances are reabsorbed into the blood and others secreted. In the collecting duct, final modification of glomerular filtrate takes place before it is excreted as urine. The glomerular filtration rate, the quantity of filtrate formed per minute, is mediated by the net effect of (1) the pressure of blood entering the glomerulus, (2) colloid osmotic pressure, and (3) filtration pressure.

When water intake is low and solute concentration consequently high, the hypothalamus discharges nerve impulses which trigger the posterior lobe of the pituitary gland to secrete antidiuretic hormone, or **ADH**. ADH acts on the kidney tubules to make them reabsorb water and thus conserve fluid. When water intake is high, the ADH mechanism is deactivated, and the kidneys excrete larger quantities of dilute urine. In diabetes insipidus, the ADH mechanism is impaired, with the result that urinary output is excessive and the patient suffers a water deficit.

Disturbances in fluid balance can result from changes either in water content or in the total quantity of electrolytes. When the kidneys retain sodium, water tends to be retained in order to balance osmotic pressure. Sodium retention is mediated by the circulating level of the hormone aldosterone. Increased aldosterone levels result in increased reabsorption of sodium from glomerular filtrate into blood. As sodium is retained, osmotic pressure of body fluids increases, and ADH secretion is triggered so as to mediate retention of water.

When a patient is dehydrated, body fluids are not only reduced but also more concentrated. When intravenous replacement is indicated for dehydration, a hypotonic solution is used. Overhydration results from increased total body water or decreased solute content. The patient shows polyuria, confusion due to cerebral edema, and nausea and vomiting because of decreased serum sodium. Under clinical circumstances, overhydration is treated by infusion of hypertonic mannitol to produce an osmotic diuresis.

Intake and output (I and O) records are kept for patients whose fluid balance requires close monitoring. Among the various intravenous solutions that may be administered, depending on the patient's individual condition, are isotonic, hypotonic, hypertonic, and balanced salt solutions. In administering an IV infusion, it is necessry to regulate flow rate as per the physician's orders. The drip rate of the IV solution is calculated by means of a formula for computing drops per minute: drops per minute equals total volume of solution times the drip rate of the tubing, divided by the number of hours times minutes per hour.

The sum total of the forces which tend to move fluid in one direction or the other across the capillary membrane is referred to collectively as capillary dynamics. Capillary dynamics includes the principles of osmosis, diffusion, and pressure gradients. Factors which tend to move fluid from serum to interstitial fluid are (1) capillary pressure, (2) interstitial fluid pressure, and (3) interstitial fluid colloid osmotic pressure. The factor which tends to move fluid in the opposite direction, from interstitial fluid to serum, is serum colloid osmotic pressure.

Capillary pressure is the pressure which remains at a capillary level from the mean arterial blood pressure imposed by the cardiovascular system.

Interstitial fluid pressure, a negative value normally, is caused by the slight suction created as lymph is drained from interstitial space and enters the one-way flow of lymphatic circulation.

Interstitial fluid colloid osmotic pressure is a result of the presence of a small amount of protein in interstitial space.

Serum colloid osmotic pressure results from the large quantity of non-diffusible protein in the blood.

When these four pressures are altered to a point at which negative interstitial fluid pressure becomes zero, fluid accumulates in interstitial spaces and a condition of edema exists.

One of the most common problems of fluid imbalance is edema. This pathological condition is associated with congestive heart failure, extensive burns, allergic reactions, malignant lesions, and dietary protein deficiencies. All these conditions have as a common denominator the collection of excess fluid in interstitial spaces. In each case, an imbalance occurs because of impairment of one or more of the forces which effect transfer of fluid across capillary membranes. A common cause of edema is increased capillary pressure, which accompanies congestive heart failure. Low serum protein—hypoproteinemia—also produces edema, as in the case of loss of serum protein from denuded areas of burned skin. Localized edema may develop from lymphatic blockage following a radical mastectomy and from increased capillary porosity due to the release of histamine in allergies.

FLUID BALANCE

QUESTION 198 The relative stability of the electrolyte composition of body fluids throughout a constant turnover of molecules and ions is termed *dynamic equilibrium/homeostasis* (circle one).

ANSWER dynamic equilibrium

QUESTION 199 Osmotic shifts are the result of *concentration/pressure* (circle one) gradients.

ANSWER concentration

QUESTION 200 In osmosis, water moves across a semipermeable membrane from the side with a *high/low* (circle one) concentration of solute to the side with a *high/low* (circle one) concentration of solute.

ANSWER low . . . high

QUESTION 201 An isotonic solution has the same _____ of particles as body fluids.

ANSWER concentration

QUESTION 202 A hypertonic solution is _____ concentrated than body fluids; a hypotonic solution, _____ concentrated than body fluids.

ANSWER more . . . less

QUESTION 203 Total body water is a balance between _____ and _____.

ANSWER intake . . . output

QUESTION 204 Output of water occurs in _____, in _____, and through the _____ and lungs.

ANSWER urine . . . feces . . . skin

SUMMARY OF UNIT TWO

When fluid intake is excessive, the kidneys remove a large quantity of _____ urine; when intake is reduced, the kidneys remove a small quantity of _____ urine.

QUESTION 205

dilute . . . concentrated

ANSWER

The structures making up the nephron are the _____, the _____, and the _____ duct.

QUESTION 206

glomerulus . . . tubule . . . collecting

ANSWER

Antidiuretic hormone, released when water intake is *high/low* (circle one), influences the kidney to *remove/reabsorb* (circle one) water.

QUESTION 207

low . . . reabsorb

ANSWER

The effect produced by aldosterone is the reabsorption of _____; this reabsorption of sodium brings about reabsorption of _____.

QUESTION 208

sodium . . . water

ANSWER

Capillary dynamics refers to the sum of _____ which tend to move _____ across the _____ membrane.

QUESTION 209

forces (pressures) . . . fluid . . . capillary

ANSWER

Forces which tend to move fluid from plasma to interstitial fluid are:

(1) _____ pressure

(2) _____ pressure

(3) interstitial fluid _____ pressure

QUESTION 210

(1) capillary

(2) interstitial fluid

(3) colloid osmotic

ANSWER

The force which tends to move fluid from interstitial fluid into serum is _____ pressure.

QUESTION 211

serum colloid osmotic

ANSWER

FLUID BALANCE

QUESTION 212 The presence of a small amount of protein in interstitial space accounts for interstitial fluid _____ _____ pressure.

ANSWER colloid osmotic

QUESTION 213 The presence of a large quantity of non-diffusible protein in the blood accounts for _____ _____ _____ pressure.

ANSWER serum colloid osmotic

QUESTION 214 Edema, or excess fluid in _____ _____, is generally caused by an imbalance in capillary _____.

ANSWER interstitial space ... dynamics

QUESTION 215 Pathological conditions accompanied by edema formation include (1) _____ _____ _____, (2) _____, (3) _____, and (4) _____.

ANSWER congestive heart failure ... burns ... allergy ... cancer

QUESTION 216 Four physiological abnormalities which are immediate causes of edema are (1) increased capillary _____, (2) increased capillary _____, (3) low serum _____, and (3) _____ blockage.

ANSWER pressure ... porosity ... protein ... lymphatic

UNIT THREE • ELECTROLYTE BALANCE

ROLE OF THE KIDNEY IN ELECTROLYTE BALANCE

ITEM 1

Unit Two dealt primarily with fluid balance. The role of the kidney was conspicuous. The present unit investigates the mechanisms for control of electrolyte balance, and the ill effects of electrolyte imbalance. Here, too, the function of the kidney is paramount. The primary focus of attention in this Unit is on the elimination or retention of specific ions in maintaining homeostasis in the various fluid compartments.

Regulation of the composition of extracellular fluids, the body's "internal environment," is indispensable to the life of the cell. To accomplish this regulatory function, the kidneys control five distinct aspects of extracellular fluid balance: (1) volume of extracellular fluid, (2) osmolarity of body fluid, (3) blood volume, (4) concentration of specific electrolytes, and (5) hydrogen ion concentration. Specifically, the kidneys control these factors, respectively, by:

(1) maintaining the optimum volume of extracellular fluids by varying the volume of urinary output,

(2) maintaining osmolarity of body fluids at a level which is isotonic with intracellular fluids,

(3) maintaining blood volume through controlled reabsorption of water,

(4) providing reabsorption or secretion of specific electrolytes in glomerular filtrate, and

(5) varying the pH (acidity) of urine to maintain a stable acid-base balance in body fluids.

QUESTION 1

The kidneys control the volume of extracellular fluid by varying the volume of _____ _____.

ANSWER

urinary output

QUESTION 2

The kidneys control the osmolarity of body fluid by varying the excretion of water so that osmolarity of body fluids is _____ with intracellular fluid.

ANSWER

isotonic

QUESTION 3

The kidneys control blood volume by controlled _____ of water.

ANSWER

reabsorption

109

110 ELECTROLYTE BALANCE

QUESTION 4 The kidneys control the concentration of electrolytes by _____ of specific electrolytes from glomerular filtrate and by _____ of other specific electrolytes into glomerular filtrate.

ANSWER reabsorption ... secretion

QUESTION 5 The kidneys control hydrogen ion concentration by varying the _____ of urine.

ANSWER pH (acidity)

ITEM 2 TRANSPORT ACROSS TUBULAR EPITHELIUM

As fluid courses through the lumen of each kidney tubule, the renal epithelium reabsorbs or secretes a variety of substances along the way. Reabsorption is the transfer of solute particles, either molecules or ions, from glomerular filtrate into blood. Substances removed from the tubular lumen enter interstitial fluid spaces surrounding the nephron and diffuse into the *peritubular capillary plexus*, shown in Figure 22. Through this plexus, reabsorbed substances return to venous circulation.

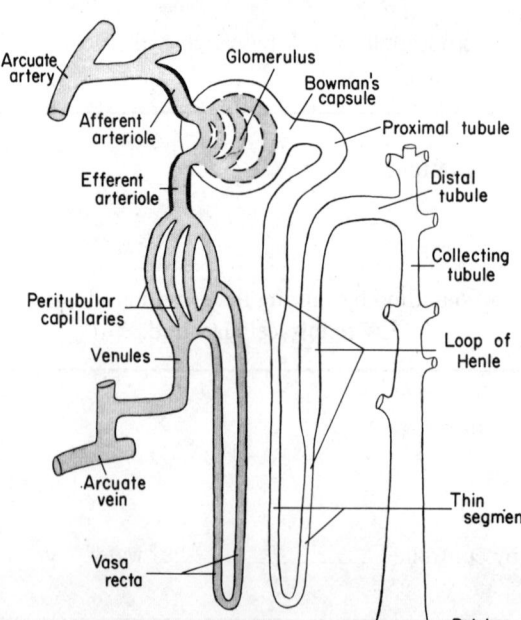

Figure 22. The functional nephron. (From Guyton: *Textbook of Medical Physiology*. 5th Ed. Philadelphia: W. B. Saunders Company, 1976.)

The direction of transport in secretion is opposite that in reabsorption. Secretion involves the transfer of substances from blood, through interstitial fluids, into glomerular filtrate and urine. When renal function is impaired, either the reabsorptive process or the secretory process may be affected.

Reabsorption of some substances and secretion of other substances occur in opposite directions across the tubular epithelium. The direction of reabsorption is from _____ to _____. — QUESTION 6

glomerular filtrate ... blood — ANSWER

Secretion involves transfer of substances from _____ capillaries, through _____ spaces, into _____. — QUESTION 7

peritubular ... interstitial ... glomerular filtrate — ANSWER

REABSORPTION IN THE PROXIMAL TUBULES — ITEM 3

Some electrolytes are reabsorbed along almost the entire length of the kidney tubules, while others are preferentially reabsorbed in certain segments only. Most reabsorption takes place in the *proximal convoluted tubules*, that is, the tubules nearest the glomerulus. For the location of the proximal tubules, look back to Figure 22.

Practically all glucose and amino acids that are present in glomerular filtrate are reabsorbed by active transport in the proximal tubules. About 80 per cent of the sodium, potassium, chloride, and other electrolytes are also reabsorbed by active transport in the proximal tubules. Active transport of sodium and other cations leaves a relative excess of negatively charged particles, which causes chloride, bicarbonate, and some phosphate to diffuse along electrical gradients. With reabsorption of solute, diffusion of water occurs by osmosis.

Reabsorption of essentially all glucose and amino acids contained in glomerular filtrate takes place in the _____ convoluted tubules. — QUESTION 8

proximal — ANSWER

Almost 80 per cent of the sodium, potassium, chloride, and other electrolytes, as well as water, is reabsorbed in the _____ tubules. — QUESTION 9

proximal convoluted — ANSWER

ELECTROLYTE BALANCE

QUESTION 10　The transport mechanism by which the renal epithelium reabsorbs molecules and ions is _____ _____.

ANSWER　active transport

QUESTION 11　Accompanying the reabsorption of particles by active transport is the reabsorption of water by _____.

ANSWER　osmosis

QUESTION 12　Chloride and bicarbonate diffuse across the tubular membrane along an _____ gradient.

ANSWER　electrical

QUESTION 13　The gradient for the diffusion of chloride and bicarbonate is created by the negativity left by the passage of positive _____ ions across the tubular membrane.

ANSWER　sodium

ITEM 4　REABSORPTION IN THE LOOP OF HENLE

Water is reabsorbed in the descending *loop of Henle* (look back to Figure 22 for the location of the loop of Henle). Because of its impermeability, the ascending loop absorbs almost no water. Sodium, however, is actively transported across the epithelium of the ascending loop. Sodium goes from glomerular filtrate in the tubular lumen, through the renal epithelium, into peritubular fluid, and, subsequently, into blood. While in the peritubular interstitial fluid, sodium increases the osmotic pressure of fluids surrounding the descending loop, and thereby increases the reabsorption of water.

QUESTION 14　Water is reabsorbed in the *ascending/descending* (circle one) loop of Henle, but not in the *ascending/descending* (circle one) loop, because of the latter's impermeability.

ANSWER　descending . . . ascending

Because of its presence in the peritubular interstitial fluid, sodium reabsorbed in the ascending loop of Henle increases the _____ concentration of the interstitial fluid, promoting the osmotic reabsorption of _____ in the descending loop.

QUESTION 15

solute ... water

ANSWER

REABSORPTION IN THE DISTAL TUBULES

ITEM 5

Final modification of glomerular filtrate takes place in the *distal convoluted tubules* and in the *collecting ducts*. (Look back to Figure 22 for the location of these parts of the nephron.) Water is reabsorbed under the influence of antidiuretic hormone, the increased secretion of which increases distal tubular permeability. Sodium is transported in varying amounts, under the control of aldosterone, from glomerular filtrate into interstitial fluid and blood. As tubular reabsorption of water increases, urinary volume decreases, and the patient becomes oliguric or anuric. When tubular reabsorption of water is lessened, polyuria ensues.

Reabsorption of sodium and water in the distal tubules is regulated, respectively, by _____ and _____ _____.

QUESTION 16

aldosterone ... antidiuretic hormone

ANSWER

When ADH secretion increases the permeability of distal tubular epithelium, urinary output tends toward *oliguria/polyuria* (circle one).

QUESTION 17

oliguria

ANSWER

SOLUTE SECRETION IN THE DISTAL TUBULES

ITEM 6

Besides reabsorbing sodium and water, the distal convoluted tubules and collecting ducts secrete hydrogen ions, potassium ions, and ammonia, removing these substances from body fluids by transporting them from interstitial fluids into the tubular lumen.

Although most of the water and solutes are reabsorbed before the glomerular filtrate

114 ELECTROLYTE BALANCE

reaches the distal convoluted tubules, the differential reabsorption of the substances which remain in glomerular filtrate at the distal segments is extremely important in regulation of fluid and electrolyte concentration of body fluids.

QUESTION 18

Substances secreted by the distal convoluted tubules and collecting ducts include _____ ions, _____ ions, and _____.

ANSWER

hydrogen ... potassium ... ammonia

QUESTION 19

Secretion transports substances from _____ to _____.

ANSWER

blood ... glomerular filtrate

ITEM 7 CONTROL OF SPECIFIC CATIONS

Appreciable variation in the concentration of certain electrolytes retained in body fluids can drastically affect a number of body functions. The cationic component of extracellular fluids, rather than the anionic component, requires precise regulation. The three specific cations in most critical need of exact concentration control are potassium, sodium, and calcium.

Lowered concentration of extracellular potassium, *hypokalemia*, causes paralysis; high serum potassium, *hyperkalemia*, brings on cardiac arrhythmias. Increased serum sodium, *hypernatremia*, leads to fluid retention and hypervolemia, imposing an extra workload on the heart. Low serum calcium, *hypocalcemia*, increases the permeability of nerve cell membranes to sodium and leads to tetany.

QUESTION 20

Regulation of the cationic concentration of body fluids is especially important because of the adverse effects of variation in the levels of _____, _____, and _____.

ANSWER

potassium ... sodium ... calcium (in any order)

The effect of hypokalemia is _____; of hyperkalemia, _____ _____.	QUESTION 21
paralysis . . . cardiac arrhythmias	ANSWER
Extreme hypocalcemia leads to _____ by increasing the permeability of nerve cell membranes to _____.	QUESTION 22
tetany . . . sodium	ANSWER
Because increased serum sodium causes water retention, hypernatremia produces _____.	QUESTION 23
hypervolemia	ANSWER

MECHANISMS REGULATING ALDOSTERONE SECRETION

ITEM 8

Quantitative control of reabsorption and secretion of electrolytes by renal epithelium is regulated by various hormones secreted by endocrine glands. The hormone regulating sodium retention is aldosterone, secreted by the adrenal cortex. Increased levels of aldosterone in the blood perfusing the kidney stimulate the enzymatic machinery responsible for transport of sodium from glomerular filtrate into peritubular capillaries. Low sodium ion concentration, or hyponatremia, triggers the secretion of aldosterone, which promotes sodium reabsorption so as to conserve total body sodium until the hyponatremic stimulus subsides.

Other bodily conditions in addition to reduced serum sodium will bring on increased aldosterone secretion. Among these entities are increased potassium concentration, decreased cardiac output, hypovolemia, and stress.

Hyponatremia stimulates the secretion of _____; secretion of this hormone will continue until sufficient _____ is reabsorbed to cause the original stimulus to subside.	QUESTION 24
aldosterone . . . sodium	ANSWER

QUESTION 25	Aldosterone secretion is increased in response to

(1) reduced _____ concentration in extracellular fluids

(2) increased _____ concentration in extracellular fluids

(3) _____ cardiac output

(4) _____

(5) _____

ANSWER

(1) sodium

(2) potassium

(3) decreased

(4) hypovolemia

(5) stress

ITEM 9 EFFECTS OF HYPERNATREMIA AND HYPONATREMIA

If aldosterone is secreted in large quantities, sodium is excessively retained. The major clinical effect of excessive sodium, or hypernatremia, is fluid retention, brought on by the increased osmotic pressure created in body fluids as a result of the excessive concentration of sodium ions. Increased blood volume due to fluid retention overworks the heart and, if prolonged enough, will eventually lead to congestive heart failure.

Hyponatremia, low sodium ion concentration, causes fluid depletion and decreased blood volume. Hyponatremia diminishes the strength of cardiac contractions.

QUESTION 26

Water is reabsorbed with sodium reabsorption as a result of increased _____ _____ in body fluids.

ANSWER

osmotic pressure

QUESTION 27

Hypernatremia has as its major clinical effect _____, which eventually imposes a work overload on the _____.

ANSWER

fluid retention ... heart

QUESTION 28

Hyponatremia reduces _____ volume and diminishes the strength of _____ _____.

ANSWER

blood (fluid) ... cardiac contractions

SODIUM DEPLETION IN ADDISON'S DISEASE
ITEM 10

Addison's disease results from atrophy of the adrenal glands, tuberculous destruction, or invasion of the adrenal cortices by cancer. With degeneration of the adrenal cortex, insufficient aldosterone secretion reduces sodium retention to such a point that the patient becomes hyponatremic. Hyponatremia results in hypovolemia and cerebral edema, owing to the osmotic shift of fluid out of the extracellular compartment and into the intracellular compartment. If untreated, the hyponatremia accompanying Addison's disease is fatal. Treatment for Addison's disease involves (1) administration of a synthetic corticosteroid to supplement the aldosterone deficiency and (2) controlled salt intake.

QUESTION 29

In Addison's disease, the adrenal cortex fails to maintain adequate circulating levels of _____, resulting in _____ depletion.

ANSWER

aldosterone ... sodium

QUESTION 30

Addison's disease brings on a *hypertonicity/hypotonicity* (circle one) of extracellular fluid leading to *hypervolemia/hypovolemia* (circle one), which is eventually fatal if untreated.

ANSWER

hypotonicity ... hypovolemia

QUESTION 31

Addison's disease is treated with a synthetic supplement for _____ and with regulated _____ intake.

ANSWER

aldosterone ... salt

SODIUM EXCESS IN CUSHING'S SYNDROME
ITEM 11

One endocrine disorder which is at the opposite extreme from Addison's disease is *Cushing's syndrome*, which develops from adrenal hypersecretion or excess secretion of adrenocorticotropic hormone (ACTH) by the anterior lobe of the pituitary gland. Several hormones, among them ACTH, stimulate the adrenal cortex to hypersecrete aldosterone, with consequent retention of sodium and water. Depending upon the extent of sodium excess, the patient with Cushing's syndrome experiences moderate to severe extracellular fluid expansion, increased blood volume, increased cardiac output, and hypertension. Because sodium reabsorption is accompanied by potassium excretion, the patient with Cushing's syndrome may also have varying degrees of hypokalemia.

QUESTION 32
Cushing's syndrome is a disorder in which there is hyperfunctioning by the _____ _____ or in which the anterior pituitary gland secretes abnormal amounts of _____ hormone.

ANSWER
adrenal cortices ... adrenocorticotropic

QUESTION 33
The adrenal cortex increases secretion of aldosterone in response to stimulation by _____ hormone.

ANSWER
adrenocorticotropic

QUESTION 34
In Cushing's syndrome, sodium excess is accompanied by water retention and all the cardiovascular consequences of increased _____ volume, including moderate to severe _____.

ANSWER
blood ... hypertension

QUESTION 35
Hypokalemia may accompany Cushing's syndrome because sodium retention is accompanied by _____ excretion.

ANSWER
potassium

ITEM 12   STRESS AND ALDOSTERONE SECRETION

Aldosterone secretion also increases in response to stress conditions such as burns, operative procedures, physical trauma, and emotional problems. The direct effect of stress is an increase in the secretion of adrenocorticotropic hormone (ACTH) by the anterior lobe of the pituitary gland. ACTH primarily stimulates the adrenal cortex to release cortisol, a glucocorticoid involved in glucose, fat, and protein metabolism and in suppressing the inflammatory process. In addition to its cortisol-releasing effect, ACTH also has a lesser, but significant, effect in stimulating aldosterone secretion. All the consequences that ensue when sodium retention follows aldosterone secretion, therefore, may follow in the wake of stress.

	QUESTION 36
Aldosterone secretion and sodium retention follow stress conditions such as _____, _____ procedures, physical _____, and _____ problems.	

	ANSWER
burns ... operative ... trauma ... emotional	

	QUESTION 37
The increased ACTH secretion that accompanies stress results in increased secretion of the anti-inflammatory hormone _____.	

	ANSWER
cortisol	

	QUESTION 38
In addition to secreting the glucocorticoid which aids in metabolism of fat, glucose, and protein, the adrenal cortex responds to ACTH stimulation by also secreting _____, which brings about sodium retention.	

	ANSWER
aldosterone	

POTASSIUM AND HYDROGEN ION EXCHANGE ITEM 13

Potassium ions are reabsorbed in the proximal tubules and secreted in the distal tubules. The potassium secretion phase—the phase affected by aldosterone—takes place as an exchange mechanism with sodium across the tubule. As circulating levels of aldosterone cause sodium to be reabsorbed from glomerular filtrate, either potassium ions or hydrogen ions are secreted in the other direction in order to maintain electrical neutrality. It is not certain what determines whether a potassium ion or a hydrogen ion will be preferentially secreted into glomerular filtrate when a sodium ion is reabsorbed. It seems plausible that when potassium ions are in excess, more of them would enter into the sodium exchange mechanism, with hydrogen ions being retained in body fluids. Conversely, when hydrogen ions exceed potassium ions in body fluids, hydrogen ions would appear to have a greater probability of being secreted in exchange for sodium, with potassium being retained. Whatever the reason for the preferential secretion of either potassium or hydrogen, it is clear that increased aldosterone secretion causes one or the other to be exchanged for reabsorbed sodium.

	QUESTION 39
As sodium ions are reabsorbed from glomerular filtrate into blood under the influence of the hormone _____, either _____ ions or _____ ions are secreted into the glomerular filtrate.	

	ANSWER
aldosterone ... potassium ... hydrogen	

ELECTROLYTE BALANCE

QUESTION 40 Preservation of _____ is responsible for the operation of the sodium exchange mechanism.

ANSWER electrical neutrality

QUESTION 41 When hydrogen ions are deficient in body fluids, the chances are that more _____ ions will be secreted in exchange for reabsorbed sodium; when potassium is deficient, it is likely that more _____ ions will be secreted.

ANSWER potassium . . . hydrogen

ITEM 14 CLINICAL CONSEQUENCES OF POTASSIUM AND HYDROGEN ION EXCHANGE

An initially hyperkalemic condition may be the stimulus for release of aldosterone to remove excess potassium in exchange for sodium. In this situation, an excess of hydrogen ions may be retained in body fluids, giving rise to a tendency to acidosis as the end result of hyperkalemia.

If an acidotic state obtains when circulating levels of aldosterone are increased, hydrogen ions are apt to be preferentially secreted in exchange for sodium. In this case, an excessive number of potassium ions may be retained, and thus acidosis may tend to promote hyperkalemia. This is an exact reversal of the situation in which hyperkalemia tends to promote acidosis.

If aldosterone causes sodium reabsorption when there is a marked deficit of both potassium and hydrogen ions, secretion appears to depend on chance availability. If potassium is secreted, there is a tendency to hypokalemia; if hydrogen ions are secreted, their loss engenders a tendency to alkalosis. Thus alkalosis may tend to promote hypokalemia, and vice versa.

QUESTION 42 Hyperkalemia tends to promote _____, while acidosis tends to promote _____.

ANSWER acidosis . . . hyperkalemia

QUESTION 43 Alkalosis tends to promote _____, while hypokalemia tends to promote _____.

ANSWER hypokalemia . . . alkalosis

ELECTRICAL PHENOMENA IN NERVE CONDUCTION

ITEM 15

Contraction of muscle depends on conduction of impulses along a nerve fiber. In the resting, or polarized, state, the total electrical charge of a nerve fiber is positive on the outside of the nerve cell membrane and negative on the inside. The voltage difference across a resting nerve cell membrane, referred to as the *resting potential*, is the result of the distribution of positive and negative ions. Sodium, a cation, is concentrated outside the cell; potassium, also a cation, is concentrated inside the cell. When sodium and potassium exchange positions across the nerve cell membrane, depolarization, or reversal of charges inside and outside the membrane, is triggered, and impulses are conducted down the length of the nerve fiber.

A resting nerve cell membrane, a polarized membrane, has an excess of positive charges

A. inside the fiber

B. outside the fiber

QUESTION 44

The correct answer is B. In the total electrolyte distribution there is an excess of positive charges on the outside of the neuron relative to the charge in the interior, making a difference in charge, the *resting membrane potential*. In the resting state, sodium is concentrated outside and potassium inside the fiber.

ANSWER

Depolarization is a reversal of charge across the membrane, triggered by influx of sodium. Depolarization and conduction of an impulse are characterized by a spread of

A. negative charges outside the fiber

B. positive charges outside the fiber

QUESTION 45

The correct answer is A. Depolarization is a wave of negativity outside the nerve cell membrane.

ANSWER

SODIUM AND POTASSIUM EXCHANGE IN NERVE CONDUCTION

ITEM 16

Conduction of an impulse depends on a change in the nerve cell membrane's permeability to sodium ions. With an appropriate stimulus, calcium ions move from their position on the cell membrane, making the pores of the membrane permeable to sodium.

122 ELECTROLYTE BALANCE

In this process of depolarization, increased permeability of the membrane to sodium ions allows sodium to flow to the inside of the nerve cell.

With inward diffusion of positive sodium ions, the charge outside the membrane reverses from positive to negative. At the same time, sodium influx creates a strong positive charge inside the membrane. Eventually, the positive charge inside the nerve cell becomes excessive, and electrical repulsion of two like charges causes potassium to move out. As this depolarization process spreads down the length of the nerve fiber, an impulse is transmitted.

To restore the nerve cell to its original polarity, the process of *repolarization* takes place. Sodium ions are moved by active transport to the outside of the membrane and potassium ions to the intracellular substance. The membrane resting potential is resumed by replacement of these ions, and the nerve is ready to conduct again.

QUESTION 46 With depolarization, the permeability of a nerve cell membrane to _____ is increased.

ANSWER sodium

QUESTION 47 As depolarization increases the influx of sodium into the nerve cell, the charge on the outside of the cell membrane changes from _____ to _____.

ANSWER positive . . . negative

QUESTION 48 Influx of sodium into the nerve cell results in an efflux of _____ out of the nerve cell.

ANSWER potassium

QUESTION 49 Repolarization, or restoration of the original ionic distribution, is accomplished by means of _____.

ANSWER active transport

ITEM 17 EFFECT OF HYPERKALEMIA ON DEPOLARIZATION

The depolarization of nerve cell membranes, fundamental to nerve conduction and muscle contraction, is immediately affected by any alteration in the potassium content of body fluids. Hyperkalemia is usually caused by the disruption of the integrity of cell membranes which accompanies tissue destruction. Disruption permits potassium, which is concentrated within cells, to leak out and thus raise its extracellular concentration. The leakage of potassium ions upsets the optimum sodium/potassium distribution, and the original voltage of the nerve cell membrane goes down. The depressed resting potential voltage brought on by hyperkalemia cuts down the intensity of sodium influx and tends to nullify the action potential and diminish nerve impulses.

Normal conduction of a nerve impulse to a muscle fiber produces vigorous contraction of muscle tissue. Hyperkalemia, by depressing the original resting potential and the voltage of the action potential, weakens muscular contraction because of the diminished nerve impulses.

Disruption of cell membranes results in leakage of _____ from _____ fluid into _____ fluid. — QUESTION 50

---

potassium ... intracellular ... extracellular — ANSWER

Hyperkalemia causes *an increase/a decrease* (circle one) in the voltage of the resting membrane potential. — QUESTION 51

---

a decrease — ANSWER

Decreased voltage in a nerve action potential affects contraction of muscle tissue as follows: — QUESTION 52

A. Fewer muscles are stimulated and fewer muscles contract.

B. The degree of contraction of individual muscle fibers is reduced.

The correct answer is A. The fault here is not in the contractile process of the muscle. The trouble lies in the weakened nerve conduction to the muscle. — ANSWER

EFFECT OF HYPERKALEMIA ON THE HEART
ITEM 18

The adverse effects of weakened muscle contraction in hyperkalemia are most dangerous in heart muscle. The myocardium becomes increasingly flaccid and dilates. Extreme hyperkalemia inhibits the normal sequence of conduction to the point of *atrioventricular block*, that is, blockage of the transmission of impulses from atria to ventricles. When this blockage occurs, ventricles contract at a much slower rate, and pumping action is correspondingly decreased. A further effect of hyperkalemia is dilatation of blood vessels. Since excess potassium ions inhibit smooth muscle activity,

ELECTROLYTE BALANCE

the vascular system loses its tonicity. Vasodilatation and cardiac flaccidity then predispose to shock. Hyperkalemia and, paradoxically, hypokalemia, by upsetting the conduction system of the heart, cause a tendency to arrhythmias. In its most extreme stage, hyperkalemia leads to cardiac arrest in diastole, as the heart becomes increasingly flaccid and finally stops.

QUESTION 53 Hyperkalemia produces vasodilatation because *an excess of/a deficiency of* (circle one) potassium ions *strengthens/inhibits* (circle one) contraction of smooth muscle.

ANSWER excess of ... inhibits

QUESTION 54 Hypokalemia, as well as hyperkalemia, tends to upset the conduction system of the heart; hypokalemia upsets the conduction system by

A. decreasing the voltage of the action potential

B. increasing the voltage of the action potential

ANSWER The correct answer is B. Although the answer to this question is not explicitly mentioned in the text of the Item, it can be figured out by reasoning that if excess potassium ions decrease voltage, then the opposite condition found in hypokalemia, a deficit of potassium ions, would increase voltage. By increasing the voltage, hypokalemia makes it more difficult for the normal threshold for contraction to be reached.

QUESTION 55 In an atrioventricular block, the heart continues to fill with blood, but it cannot pump effectively because _____ are not transmitted from _____ to _____ .

ANSWER impulses ... atria ... ventricles

QUESTION 56 Cardiac arrest in hyperkalemia occurs in the *diastolic/systolic* (circle one) phase.

ANSWER diastolic

QUESTION 57 The flaccidity of myocardium brought on by hyperkalemia is due to weakened muscular contraction, resulting from

A. weakened impulses which inadequately stimulate heart muscle

B. weakened heart muscle which responds inadequately to stimulation

ANSWER The correct answer is A. Hyperkalemia does nothing to weaken muscle; it diminishes the number of the impulses transmitted along the nerve fiber to the muscle.

INTRAVENOUS INFUSIONS IN HYPERKALEMIA AND HYPOKALEMIA

ITEM 19

Because of the potentially lethal effects of hypokalemia and hyperkalemia, potassium additives to intravenous fluids are very carefully regulated. Since potassium is poorly conserved in the body, dangerously low levels may develop within a few days in patients who are unable to take food and fluids by mouth and who rely entirely on IV fluids for electrolytes. In these patients, potassium must be included in the solution. On the other hand, addition of excess potassium can be very dangerous, particularly in a patient with impaired kidney function and unreliable excretion. In most situations, it is recommended that no more than 40 mEq. of potassium be infused in an 8-hour period, with this amount always diluted in a liter of fluid. A drip infusion must be employed, never an injection or intravenous push.

Regulation of intravenous infusion of potassium generally stipulates not more than _____ mEq. of potassium added to _____ cc. of fluid, to run for a period of _____ hours.

QUESTION 58

---

40 ... 1000 ... 8

ANSWER

The danger of excess infusion of potassium is especially critical in patients with _____ disorders.

QUESTION 59

---

kidney

ANSWER

REGULATION OF SERUM CALCIUM LEVELS

ITEM 20

Two hormones with opposite effects regulate serum calcium levels. *Parathyroid hormone*, secreted by the parathyroid glands, tends to protect against calcium deficiency, or *hypocalcemia*. Calcitonin, secreted by the thyroid gland, protects against calcium excess, or *hypercalcemia*.

Parathyroid hormone, secreted in response to low serum calcium concentration, causes the dissolution and reabsorption of the mineral content of bone. Specifically, parathyroid hormone effects demineralization by increasing the activity of *osteoclasts*, the bone cells which effect bone dissolution. Increased osteoclastic activity releases calcium and phosphorus into extracellular fluid, thus raising calcium levels. Increased serum calcium in turn inhibits further parathyroid release, because the original stimulus for its secretion is removed.

Calcitonin, which acts considerably more rapidly than parathyroid hormone, protects against hypercalcemia by blocking reabsorption of calcium from bone.

QUESTION 60

Parathyroid hormone secretion is stimulated by _____ serum calcium levels.

ANSWER

low

QUESTION 61

Reabsorption of calcium is blocked by the hormone _____, secreted by the _____ gland.

ANSWER

calcitonin . . . thyroid

QUESTION 62

Serum calcium levels are increased as parathyroid hormone promotes _____ of bone.

ANSWER

demineralization (or resorption)

QUESTION 63

Parathyroid hormone protects against _____ by stimulating _____ to dissolve bone, making calcium available for reabsorption into extracellular fluid.

ANSWER

hypocalcemia . . . osteoclasts

ITEM 21 PARATHYROID AND GASTROINTESTINAL ABSORPTION OF CALCIUM

Another effect of parathyroid hormone is to increase the amount of calcium and phosphorus absorbed from the gastrointestinal tract. This effect is dependent upon an adequate supply of vitamin D.

When vitamin D is present in optimum concentration, the parathyroid glands control the gastrointestinal absorption in response to the circulating level of calcium ions in body fluid. Decreased calcium causes increased secretion of parathyroid hormone, which in turn increases serum calcium by promoting bone dissolution and absorption of calcium by the intestinal epithelium.

Parathyroid hormone increases the gastrointestinal absorption of _____ and _____.	QUESTION 64
calcium ... phosphorus	ANSWER
Parathyroid control of gastrointestinal absorption is critically dependent upon a sufficient amount of _____ in the body.	QUESTION 65
vitamin D	ANSWER

PARATHYROID AND RENAL CONSERVATION OF CALCIUM

ITEM 22

A further effect of parathyroid hormone is the augmentation of renal reabsorption of calcium. About 1/8 of the daily intake of calcium is normally excreted in urine. Increased parathyroid activity recoups some of this calcium by tubular transport from the filtrate back into extracellular fluid.

Parathyroid hormone also causes the kidney to increase its excretion of phosphate, leaving excess calcium ions in extracellular fluid.

Parathyroid hormone causes the kidney to increase excretion of _____, leaving an excess of _____ in extracellular fluids.	QUESTION 66
phosphate ... calcium	ANSWER
Secretion of parathyroid hormone promotes reabsorption of calcium by tubular transport from glomerular filtrate back into _____ fluid.	QUESTION 67
extracellular	ANSWER

ITEM 23 — EFFECTS OF ABNORMAL PARATHYROID SECRETION

Deficiency of parathyroid hormone, a condition known as hypoparathyroidism, depletes serum calcium levels; excess parathyroid hormone, hyperparathyroidism, elevates them. The effects of too high or too low a concentration of calcium ions in the extracellular fluid are drastic and quick. Hypoparathyroidism causes a generalized muscular spasm called *tetany* which can bring on death by respiratory paralysis. Hyperparathyroidism weakens the structure of bone and depresses the nervous system.

QUESTION 68

Low calcium tetany develops in

A. hypoparathyroidism

B. hyperparathyroidism

ANSWER

A. This answer is correct. Low calcium tetany develops when serum calcium levels drop to about two-thirds their normal value. Decreased calcium increases permeability of nerve cell membranes. Increased permeability to sodium lowers the membrane potential and depolarizes the nerve fiber, causing it to discharge spontaneously. Repetitive nerve discharge causes repetitive contraction of skeletal muscle fibers, with no interval for relaxation of these muscles. Continued rapid nerve-muscle firing at a rate which allows no relaxation constitutes tetany.

B. This answer is incorrect. Hyperparathyroidism produces higher serum calcium levels. High serum calcium level decreases the permeability of cell membranes, thus preventing depolarization and excitation of nerve fibers. In other words, hyperparathyroidism causes hypercalcemia; hypercalcemia causes a depression of the nervous system which is opposite to the excitation characteristic of tetany.

ITEM 24 — EFFECTS OF HYPOPARATHYROIDISM

Parathyroid hormone normally maintains the serum calcium level at a relatively fixed value. Ionized calcium in body fluids regulates permeability of cell membranes. When serum calcium is low, as it is with hypoparathyroidism, the permeability of the membrane of nerve cells increases. This increased permeability permits a leakage of sodium and potassium across the membrane, resulting in unwanted and uncontrollable *depolarization*. Continued repetitive depolarization of the nerves to the skeletal muscles causes these muscles to go into *tetanic contractions* (rapid, uncontrollable spasms). Tetany leads to death by spasm of the respiratory muscles. Serum calcium concentration needs to be only 30 per cent below normal for tetany to appear.

Abnormally low serum calcium levels result in a clinical condition called tetany, especially lethal when it goes into spasm of the _____ muscles.	QUESTION 69
respiratory	ANSWER
Ionized calcium in body fluids affects conduction along nerve fibers by regulating the _____ of the nerve cell membrane.	QUESTION 70
permeability	ANSWER
Hypocalcemia specifically increases the permeability of neurons to _____, which results in spontaneous and repetitive _____ of these nerve cell membranes.	QUESTION 71
sodium . . . depolarization	ANSWER
Muscle tissue responds to uncontrolled stimulation by repetitive _____, which is, in essence, tetany.	QUESTION 72
contraction	ANSWER

EFFECTS OF HYPERPARATHYROIDISM

ITEM 25

Hyperparathyroidism occasionally results from a tumor of one of the parathyroid glands. In this case, osteoclastic activity removes an abnormal amount of calcium and phosphorus from bone storage, elevating serum calcium levels and causing increased excretion of phosphorus. Decalcification areas develop as small holes in bone tissue. Osseous tissue is weakened and is vulnerable to frequent fractures.

Elevated serum calcium has a second pathological effect. It decreases the permeability of nerve cell membranes to sodium, thus affecting the depolarization process. Less sodium enters the cell, rendering the membrane electrically inactive and decreasing the excitability of nerve fibers. This is a situation exactly opposite to the tetanic condition of over-excitable nerve fibers. The consequences of the inactivated membrane of the nerve cell are depression of the central and peripheral nervous system, sluggishness, constipation, and lack of appetite.

ELECTROLYTE BALANCE

QUESTION 73 Small holes in bone tissue develop in hyperparathyroidism as a result of decalcification aimed at raising serum _____ levels.

ANSWER calcium

QUESTION 74 Hypercalcemia results in a _____ in the permeability of nerve cell membranes.

ANSWER decrease

QUESTION 75 Decreased permeability of nerve cell membranes to sodium entrance makes the membrane electrically _____ and brings on a _____ in excitability of nerve fibers.

ANSWER inactive ... decrease

QUESTION 76 Decreased excitability of nerve fibers results in _____ of the nervous system.

ANSWER depression

ITEM 26 — EFFECT OF CALCIUM IONS ON MUSCLE TISSUE

Hypercalcemia can occur as a result of an accidental infusion of excessive calcium ions, as one consequence of the release of calcium in metastatic bone disease, or as a concomitant of other conditions. The effect of hypercalcemia on cardiac muscle is the opposite of the effect of hyperkalemia. While high serum potassium levels cause cardiac muscle flaccidity and diminished contractile strength, excess calcium increases the contractile force of heart muscle, leading to spasticity.

The effects of hypocalcemia are similar to those of hyperkalemia. Hypocalcemia triggers spontaneous depolarization of nerve tissue. Depolarization then leads to cardiac arrhythmias and decreased contractile strength.

The effect of hypocalcemia on the muscles of respiration is especially acute. Indeed, hypocalcemia is likely to kill the patient by producing tetany of the respiratory muscles long before the heart muscle has had time to be affected lethally.

The effects of hypercalcemia and hyperkalemia on contraction of the myocardium are A. the same B. the opposite	QUESTION 77
The correct answer is B. Hypercalcemia has an effect on heart muscle opposite to that of hyperkalemia. Hypocalcemia, on the other hand, affects heart muscle in the same way as hyperkalemia, causing flaccidity, weakened contractile strength, and so on. Hypercalcemia increases the contractile force of muscle.	ANSWER
The most serious effect of hypocalcemia, clinically, is A. respiratory arrest in tetany B. cardiac flaccidity	QUESTION 78
The correct answer is A. In hypocalcemia, the respiratory paralysis which follows respiratory muscle spasm, or tetany, brings on death before the hypocalcemia has seriously affected the heart muscle.	ANSWER

THE INTERRELATIONSHIP OF ELECTROLYTE AND FLUID PROBLEMS

ITEM 27

The preceding Items in this Unit dealt with the physiological mechanisms which regulate electrolyte concentrations in body fluids. The remaining Items will describe common clinical abnormalities, including excesses and deficiencies, in sodium, potassium, and calcium levels.

Many fluid and electrolyte problems are multifaceted and interrelated. For example, an abnormality in total body water is reflected in serum electrolyte levels and hematocrit. An excess or deficit in sodium ions, on the other hand, alters extracellular and intracellular fluid volume. The interrelationship of electrolyte and fluid problems is also apparent in the case of reabsorption of sodium in renal tubules in exchange for potassium or hydrogen ion secretion; an excess or deficiency in any one of the three ions involved in the exchange will result in an adjustment in the circulating levels of the other two. While each of the Items that follow will take some original imbalance as its point of departure, the reader should bear in mind that any single electrolyte problem treated in an Item may have further implications in regard to the adjustments that are initiated in other ions as homeostasis is regained or lost.

The interrelationship of fluid and electrolyte problems implies that correction of one imbalance involves adjustment of both _____ concentrations and _____ balance.	QUESTION 79
electrolyte ... fluid	ANSWER

ELECTROLYTE BALANCE

QUESTION 80
The rationale for therapeutic intervention in electrolyte and fluid imbalances is the need for maintenance or restoration of _____.

ANSWER
homeostasis

ITEM 28 SODIUM LEVELS AND FLUID BALANCES

Because sodium is the principal cation in extracellular fluids and has the potential for altering osmotic shifts and blood volume, sodium levels and fluid balance are always closely interrelated. Indeed, the whole discussion of isotonic, hypotonic, and hypertonic fluid shifts in Unit Two may be considered in terms of sodium levels.

QUESTION 81
The predominant positive ion present in ECF is _____.

ANSWER
sodium

QUESTION 82
Sodium levels always have implications for

A. levels of other ions
B. fluid balance

ANSWER
The correct answer is B. While it is true that a change in sodium levels may initiate adjustment of levels of other ions, the factor which is invariably influenced by sodium levels is fluid balance.

ITEM 29 SIGNS AND SYMPTOMS OF HYPERNATREMIA

The initial signs and symptoms of uncompensated hypernatremia are the same as those of dehydration or hyperosmolarity. Blood viscosity is increased. The patient may be thirsty as a result of the hypertonicity of fluids perfusing the hypothalamus. Urinary output diminishes as the kidney conserves water. Mucous membranes are dry and skin turgor poor because of relative fluid deficit. Unavailability of fluid sufficient for adequate perspiration may cause the patient's temperature to rise. Finally, the patient may become restless, apprehensive, depressed, or comatose as fluid shifts from brain cells to compensate for the initial hyperosmolarity of extracellular fluids. Treatment of hypernatremia involves fluid replacement, either by mouth or intravenously with a hypotonic solution, until the kidney can adjust the sodium and water balance. Manifestations of hypernatremia may rapidly progress to hypervolemia and hypertension as the excess sodium begins to cause water reabsorption and expansion of the extracellular compartment.

The initial signs of uncompensated hypernatremia are similar to _____. **QUESTION 83**

dehydration (hyperosmolarity) — ANSWER

Water deficit in uncompensated hypernatremia causes skin and mucous membranes to be _____ and, because perspiration is inadequate, may cause body temperature to be _____. **QUESTION 84**

dry ... elevated — ANSWER

In terms of urinary output, the hypernatremic patient is *oliguric/polyuric* (circle one). **QUESTION 85**

oliguric — ANSWER

Mental symptoms observed in the hypernatremic patient are _____ and _____, which, in extreme cases, may lead to _____. **QUESTION 86**

apprehension (restlessness) ... depression ... coma — ANSWER

SIGNS AND SYMPTOMS OF HYPONATREMIA
ITEM 30

Signs and symptoms of hyponatremia are identical to those of water intoxication or hypo-osmolarity of body fluids. Serum sodium levels are depressed. The patient may have a headache, become restless and disoriented, and go into convulsions as a consequence of cerebral edema as brain cells share the water excess. Cerebral edema also may stimulate the emetic center in the medulla and bring on nausea and vomiting. Muscular weakness follows alterations in the normal distribution of sodium involved in the depolarization process. If renal function is normal, the patient will show polyuria until antidiuretic hormone and aldosterone can return body fluids to normalcy. When hyponatremia requires clinical treatment, hypertonic mannitol is infused to promote osmotic diuresis.

QUESTION 87
Cerebral symptoms of hyponatremia include headache, _____, _____, and _____.

ANSWER
restlessness . . . disorientation . . . convulsions

QUESTION 88
Gastrointestinal problems associated with hyponatremia, which include _____ and _____, stem ultimately from cerebral _____.

ANSWER
nausea . . . vomiting . . . edema

QUESTION 89
In the hyponatremic patient, muscle contraction becomes weak because of alterations in the ionic ratio involved in _____ of nerve cell membrane.

ANSWER
depolarization

QUESTION 90
The hyponatremic patient with unimpaired renal function will develop *oliguria/polyuria* (circle one).

ANSWER
polyuria

ITEM 31   REGULATION OF POTASSIUM BALANCE

Regulation of potassium is extremely important because any imbalance in the levels of this ion is potentially lethal. Potassium is the principal cation in intracellular fluid and is a contributor to intracellular osmolarity. This ion also enters into the depolarization and repolarization of nerve tissue as well as skeletal, smooth, and cardiac muscle.

Potassium concentration within intracellular fluids is lost in metabolic disturbances, disease processes, or trauma which interrupts the integrity of the cell membrane. When intracellular potassium is lost, extracellular levels, normally 3.8 to 4.5 mEq./L., rise. Sodium and hydrogen ions enter the cell to maintain electrical neutrality. When cellular integrity and anabolic metabolism are restored, potassium will reenter the cell. Potassium can also enter the cell in large quantity when insulin promotes glucose entry.

The functions of potassium within cells include

(1) contributing to intracellular _____ and

(2) participating in depolarization and repolarization of _____ tissue and of _____ muscle, _____ muscle, and _____ muscle.

QUESTION 91

(1) osmolarity

(2) nerve ... skeletal ... smooth ... cardiac

ANSWER

Serum potassium levels are normally _____ to _____ mEq./L.

QUESTION 92

3.8 ... 4.5

ANSWER

With disruption of cell membranes, potassium is released from _____ fluids into _____ fluids.

QUESTION 93

intracellular ... extracellular

ANSWER

When potassium is lost from intracellular fluids, the resultant electronegativity attracts _____ and _____ ions, which enter cells.

QUESTION 94

sodium ... hydrogen

ANSWER

Intracellular potassium levels are restored by _____ metabolism and administration of insulin, which promotes _____ entry into cells.

QUESTION 95

anabolic ... glucose

ANSWER

CAUSES OF HYPERKALEMIA

ITEM 32

The major causes of hyperkalemia stem from excessive release of potassium from within cells coupled with an inability of the kidney to excrete the excess. Patients with extensive tissue damage such as occurs with crushing injuries, burns, or myocardial damage are faced with a sudden release of intracellular potassium into extracellular fluids. Excessively high potassium levels could result also from overzealous use of potassium replacement in intravenous fluids, administered either too rapidly or in excess of the recommended 120 mEq. per day.

Normally, the kidneys secrete potassium or hydrogen ions into glomerular filtrate in exchange for reabsorbed sodium ions. If renal complications cause this exchange mechanism to be impaired, dangerously high potassium levels can build up. The lethal aspects of hyperkalemia are arrhythmias, flaccidity of heart muscle, and bradycardia, possibly leading to cardiac arrest in diastole.

QUESTION 96 Potassium is excreted by the _____ in an _____ mechanism.

ANSWER kidney ... exchange

QUESTION 97 If sodium reabsorption in the kidney tubules is impaired, secretion of _____ and _____ _____ is accordingly impaired.

ANSWER potassium ... hydrogen ions (in either order)

QUESTION 98 Causes of hyperkalemia include loss of intracellular potassium from (1) _____ injuries, (2) _____, and (3) _____ damage.

ANSWER crushing ... burns ... myocardial

QUESTION 99 Hyperkalemia, in addition to resulting from release of intracellular potassium, can be caused simply by excessive intake of potassium through _____ therapy.

ANSWER intravenous

QUESTION 100 Effects of hyperkalemia on the heart include (1) _____, (2) muscle _____, (3) _____, and (4) cardiac arrest in _____.

ANSWER arrhythmias ... flaccidity ... bradycardia ... diastole

SIGNS AND SYMPTOMS OF HYPERKALEMIA

ITEM 33

The signs and symptoms of hyperkalemia appear primarily in nerve and muscle function. Gastrointestinal disturbances and diarrhea due to neuromuscular irritability are symptoms of slight hyperkalemia. In severe hyperkalemia, skeletal muscle weakness and flaccid paralysis due to neuromuscular depression are noted. Other signs of severe hyperkalemia include myocardial flaccidity, bradycardia, arrhythmias, and possible cardiac arrest in diastole, all attributable to impaired conduction through the Purkinje system and myocardium. Oliguria, anuria, and renal shutdown due to shock may accompany severe tissue injury. In hyperkalemia, laboratory tests show K^+ levels greater than 5 mEq./L.

QUESTION 101
Adverse effects of hyperkalemia are manifested primarily in _____ and _____ function.

ANSWER
nerve ... muscle (in either order)

QUESTION 102
Neuromuscular irritability which results from slight hyperkalemia causes gastrointestinal disturbances and _____.

ANSWER
diarrhea

QUESTION 103
Neuromuscular depression resulting from severe hyperkalemia is indicated by weakness of _____ muscle and _____ paralysis.

ANSWER
skeletal ... flaccid

QUESTION 104
Impaired conduction through the Purkinje system and myocardium leads to myocardial _____, _____, and cardiac _____.

ANSWER
flaccidity ... bradycardia ... arrest

QUESTION 105
The renal response to severe hyperkalemia induces _____ and _____, culminating in _____.

ANSWER
oliguria ... anuria ... renal shutdown

138 ELECTROLYTE BALANCE

ITEM 34 TREATMENT OF HYPERKALEMIA

Treatment of hyperkalemia is aimed at an immediate reduction of potassium levels within extracellular fluids. Sources of high potassium intake are immediately eliminated. Ion exchange resins such as Kayexalate, given either by mouth or by enema, absorb potassium in exchange for sodium and thus reduce total body potassium levels. Urinary excretion of potassium may be increased with diuretics such as furosemide or the thiazides. In cases of hyperkalemia due to severe renal failure, peritoneal dialysis or the artificial kidney may be employed. Potassium reentry into cells is achieved by administration of glucose and insulin to promote cellular uptake, thereby quickly reducing dangerous extracellular levels until total body potassium levels can be normalized.

QUESTION 106 Hyperkalemia is treated by (1) elimination of potassium _____, (2) use of ion exchange _____, and (3) use of diuretics such as _____ or the _____.

ANSWER intake ... resins ... furosemide ... thiazides

QUESTION 107 Rapid reentry of potassium into cells is effected by giving _____ and _____.

ANSWER glucose ... insulin

QUESTION 108 Extremely high levels of hyperkalemia can be alleviated by _____ dialysis or the _____.

ANSWER peritoneal ... artificial kidney

ITEM 35 CAUSES OF HYPOKALEMIA

Hypokalemia usually stems from inadequate potassium intake and excessive potassium loss. Inadequate intake can result from a diet deficient in potassium-rich foods, primarily fruits; from parenteral fluids that are deficient in potassium; and from anorexia and nausea. Potassium can be lost in gastrointestinal secretions; from diarrhea or fistulas; from reduced renal absorption; and in the anabolic phase following severe injury. Since potassium is found in gastrointestinal secretions, substantial loss can occur with prolonged nausea and vomiting, gastric suction, and excessive use of enemas.

The following therapeutic treatment may also cause potassium depletion: (1) prolonged use of thiazide diuretics which promote potassium excretion, (2) adrenal steroid therapy which promotes sodium reabsorption and potassium secretion across renal tubular epithelium, and (3) intravenous therapy without potassium replacement.

Inadequate potassium intake can result from (1) _____ deficiency, (2) anorexia and _____, and (3) deficient _____ fluids.	QUESTION 109
dietary . . . nausea . . . parenteral	ANSWER
Potassium is lost with nausea and _____, gastric _____, and frequent _____.	QUESTION 110
vomiting . . . suction . . . enemas	ANSWER
Drug-induced hypokalemia stems from _____ diuretics, _____ therapy, and deficient _____ therapy.	QUESTION 111
thiazide . . . adrenal steroid . . . intravenous	ANSWER

SIGNS AND SYMPTOMS OF HYPOKALEMIA

ITEM 36

Signs and symptoms of potassium deficit are reflected in altered cellular metabolism and in altered function of the neuromuscular, cardiovascular, and gastrointestinal systems. Paradoxically, many of the cardiovascular symptoms of hypokalemia are similar to those of hyperkalemia. Both excesses of and deficits in potassium produce cardiovascular aberrations, because both disturb the resting membrane potential and the action potential of nerve fibers which conduct impulses to heart muscle.

Reduced neuromuscular functioning in hypokalemia results in anorexia, weakness, loss of muscle tone and contour, flaccid paralysis, and shallow respiration. Arrhythmias, congestive heart failure, and heart block, all leading to cardiac arrest, follow upon fibrosis and necrosis of the myocardium. Smooth muscle atony brings on decreased vascular tone and hypotension. Atony of smooth muscle of the gastrointestinal tract decreases peristalsis and causes abdominal distention. Acid-base disturbances in hypokalemia are due to replacement of intracellular potassium with hydrogen and sodium ions. Laboratory tests show K^+ values in the hypokalemic patient below 3.0 mEq./L.

140 ELECTROLYTE BALANCE

QUESTION 112 Hypokalemia is accompanied by disturbances in cellular _____.

ANSWER metabolism

QUESTION 113 Three systems impaired by hypokalemia are the _____, the _____, and the _____ systems.

ANSWER neuromuscular ... cardiovascular ... gastrointestinal

QUESTION 114 Besides leading to weakness and decreased muscle tone, neuromuscular inhibition associated with hypokalemia results in _____, flaccid _____, and shallow _____.

ANSWER anorexia ... paralysis ... respirations

QUESTION 115 Symptoms of fibrosis of the myocardium which signal impending cardiac arrest are _____, congestive _____ _____, and heart _____.

ANSWER arrhythmias ... heart failure ... block

QUESTION 116 A symptom of atony of smooth muscle of the vascular system is _____; signs of atony of smooth muscle of the gastrointestinal system are decreased _____ and _____ of the abdomen.

ANSWER hypotension ... peristalsis ... distention

QUESTION 117 Potassium and hydrogen ion exchange across cell membranes, which occurs in hypokalemia, disturbs _____ balance.

ANSWER acid-base

ITEM 37 TREATMENT OF HYPOKALEMIA

Treatment of hypokalemia requires extremely cautious potassium replacement. Extracellular potassium levels are an unreliable index of intracellular potassium; therefore, replacement must be carefully geared to intracellular restoration without a rebound hyperkalemia. The safest method is to have the patient increase his or her intake of potassium-rich foods, primarily fruit, or to provide oral medication such as potassium triplex. Any patient receiving potassium medications should be watched carefully for oliguria, since renal function must be ensured in order to prevent a dangerous overload.

Extracellular potassium levels are *a reliable/an unreliable* (circle one) index of intracellular potassium. QUESTION 118

---

an unreliable ANSWER

The simplest and safest way to increase serum potassium levels is to increase dietary intake of _____. QUESTION 119

---

fruit ANSWER

Hypokalemia may revert rapidly to _____ in a patient on oral potassium medications if _____ function is impaired. QUESTION 120

---

hyperkalemia . . . renal (kidney) ANSWER

INTRAVENOUS POTASSIUM REPLACEMENT

ITEM 38

When replacement potassium is administered intravenously, extreme caution must be employed in order to prevent a rebound overload. Potassium salts are always administered in diluted form, either directly into a vein or into a muscle, and *never* in concentrated form. A common dilution is 40 mEq. to 1 liter of solution. In acute cases, such as arise in the patient with extensive burns, the drip rate is regulated so that no more than 20 mEq. of potassium is given per hour. In such acute cases, KCl is given rapidly *only* with constant EKG monitoring to assess any potassium excess. At a dilution of 40 mEq. K^+ per liter of fluid, urine flow should be at least 1 ml./minute. Pain that may develop along the vein during K^+ administration can be controlled by slowing the drip rate or by addition of a small amount of prednisone to the solution.

Intravenous potassium is always _____, commonly to _____ mEq. per liter of IV solution. QUESTION 121

---

diluted . . . 40 ANSWER

ELECTROLYTE BALANCE

QUESTION 122
In an acute situation such as extensive _____, a more rapid drip rate of potassium chloride may be ordered, but the patient's cardiovascular status must be constantly monitored with the _____.

ANSWER
burns ... EKG

QUESTION 123
When potassium is administered intravenously at 40 mEq./L., urinary output should be at least _____.

ANSWER
1 ml./min.

QUESTION 124
Pain accompanying the infusion of KCl can be alleviated by _____ the drip rate or by adding _____ to the solution.

ANSWER
slowing ... prednisone

ITEM 39 HYPOKALEMIA AND DIGITALIS TOXICITY

Hypokalemia can produce a toxic reaction to digitalis, for the drug's effects are potentiated by low serum levels of potassium. A digitalized congestive heart failure patient may also be receiving a thiazide diuretic to promote K^+ excretion. Under these circumstances, a dose of digitalis, which is not in itself toxic, may produce toxicity. Signs of digitalis toxicity include nausea, vomiting, and arrhythmias leading to heart block. Treatment consists of replacing the potassium until serum levels are normal.

QUESTION 125
The effects of digitalis are potentiated by _____.

ANSWER
hypokalemia

QUESTION 126
Congestive heart failure patients on thiazide diuretics are carefully monitored for digitalis toxicity because _____ promote _____ excretion.

ANSWER
thiazides ... potassium

Signs and symptoms of digitalis toxicity include _____, _____, and _____.	QUESTION 127
nausea ... vomiting ... arrhythmias (in any order)	ANSWER

HYPERCALCEMIA ITEM 40

Calcium ions are involved in the blood clotting mechanism, neuromuscular sensitivity, capillary permeability, and maintenance of the structure of bones and teeth. For obscure reasons, serum calcium levels are inversely related to serum phosphorus levels. Hypercalcemia occurs under conditions which promote excess mobilization of calcium from bone. Such conditions include immobility, hyperparathyroidism, and malignant metastasis to bone. Hypercalcemia also occurs when renal excretion of calcium is decreased.

Calcium ions participate in (1) blood _____ (2) neuromuscular _____ (3) capillary _____ (4) maintaining structure of _____ and _____	QUESTION 128
clotting ... sensitivity ... permeability ... bones ... teeth	ANSWER
Serum levels of calcium and _____ are inversely related.	QUESTION 129
phosphorus	ANSWER
Serum calcium levels may become elevated in such conditions as _____, _____, and _____.	QUESTION 130
immobility ... hyperparathyroidism ... metastasis to bone	ANSWER

144 ELECTROLYTE BALANCE

ITEM 41 SIGNS AND SYMPTOMS OF HYPERCALCEMIA

Signs and symptoms of hypercalcemia are manifested in the neuromuscular system, the kidneys, the gastrointestinal system, and bone. Kidney stones result from hypercalcemia, and, along with causing flank pain, may lead to renal failure. Decalcification of bone may bring on osteoporosis and pathologic fractures. Nausea and vomiting, as well as diarrhea or constipation, reflect hypercalcemic interference with gastrointestinal reflexes at the autonomic level. Lethargy, muscle weakness, confusion, irritability, and behavioral changes are signs of neurological malfunctioning in hypercalcemia. Laboratory tests show serum calcium levels of greater than 5.8 mEq./L. in the hypercalcemic patient.

QUESTION 131
Excess calcium in renal blood can cause kidney _____ and _____ pain.

ANSWER
stones ... flank

QUESTION 132
Decalcification of bones in hypercalcemia leads to _____ and pathologic _____.

ANSWER
osteoporosis ... fractures

QUESTION 133
Hypercalcemic impairment of gastrointestinal reflexes can result in nausea and _____, and in _____ or _____ as well.

ANSWER
vomiting ... diarrhea ... constipation

QUESTION 134
Besides lethargy and muscular weakness, other symptoms of the effect of hypercalcemia on the nervous system include _____, _____, and _____ changes.

ANSWER
confusion ... irritability ... behavioral

TREATMENT OF HYPERCALCEMIA

ITEM 42

Treatment of hypercalcemia includes cessation of calcium intake, administration of steroids, and increased oral or intravenous fluid intake to promote renal excretion. In emergency situations—when serum $Ca^{"}$ levels are high enough to produce EKG changes—excess calcium can be driven into bone with calcitonin, the hormone which promotes bone deposition of calcium.

In mild cases of hypercalcemia, excretion can be promoted by increasing _____ intake.

QUESTION 135

---

fluid

ANSWER

With dangerous levels of hypercalcemia, the hormone _____ may be administered to promote bone _____ of calcium.

QUESTION 136

---

calcitonin ... deposition

ANSWER

HYPOCALCEMIA

ITEM 43

Hypocalcemia occurs because of hypoparathyroidism, accidental surgical removal of the parathyroid glands during thyroidectomy, or excessive loss of calcium in pancreatitis. Hypocalcemia may also occur during pregnancy and lactation, when calcium requirements are high.

Severe hypocalcemia results in tetany and respiratory arrest. Trousseau's sign and Chvostek's sign are diagnostic tests for impending tetany. Trousseau's sign is elicited by inflating a blood pressure cuff around the arm sufficiently to stop circulation for two or three minutes. If fingers and hands go into flexion, sufficient neuromuscular irritability is present to be termed tetany. Chvostek's sign is elicited by tapping the face at the point just under the temporomandibular joint where the facial nerve crosses the neck of the mandible. If tetany is present, muscles of the face contract unilaterally.

ELECTROLYTE BALANCE

QUESTION 137

The most common conditions which bring on hypocalcemia are the following:
(1) _____
(2) _____ excision of the parathyroid glands
(3) _____
(4) pregnancy and _____

ANSWER hypoparathyroidism ... surgical ... pancreatitis ... lactation

QUESTION 138

Among the most serious effects of hypocalcemia are _____ and respiratory _____.

ANSWER tetany ... arrest

QUESTION 139

Trousseau's sign of impending tetany, as elicited in hypocalcemia, involves _____ of the hands and fingers after inflation of a blood pressure cuff around the arm.

ANSWER flexion

QUESTION 140

Chvostek's sign of impending tetany involves spasm of _____ _____ on tapping the facial nerve.

ANSWER facial muscles

ITEM 44 — SIGNS AND SYMPTOMS OF HYPOCALCEMIA

Neuromuscular signs of hypocalcemia include (1) muscle spasms which result from hyperirritability and which lead to tetany; (2) convulsions due to facilitated neuronal activity; and (3) tingling and numbness of fingers due to vascular spasms accompanying the hyperirritability. Other symptoms of hypocalcemia include specific EKG changes, palpitations, and arrhythmias due to myocardial hyperirritability. In hypocalcemia, laboratory tests show serum Ca^{++} below 4.5 mEq./L.

QUESTION 141

Because of increased neuromuscular irritability, a hypocalcemic patient is apt to show _____ _____.

ANSWER muscle spasms

Because of vascular spasms, the hypocalcemic patient may experience _____ and _____ of the fingers.	QUESTION 142
tingling . . . numbness	ANSWER
In addition to arrhythmias, cardiovascular symptoms of hypocalcemia include _____ changes and _____.	QUESTION 143
EKG . . . palpitations	ANSWER
In hypocalcemia, laboratory values for serum calcium levels are below _____ mEq./L.	QUESTION 144
4.5	ANSWER

TREATMENT OF HYPOCALCEMIA

ITEM 45

Treatment of hypocalcemia is aimed at restoring serum calcium levels. Calcium lactate, calcium chloride, or calcium gluconate may be administered orally or intravenously. Some precautions need to be observed in the intravenous infusion of calcium. Infiltration should be avoided, if possible, because of the risk of tissue sloughs. To avoid precipitation, calcium should not be added to solutions containing carbonate or phosphate. Since calcium potentiates the effect of digitalis, leading to digitalis toxicity, caution must be exercised in administering Ca^{++} to digitalized patients. Extreme caution, finally, is required with IV infusion of Ca^{++} if rebound hypercalcemia, which can result in cardiac arrest in systole, is to be avoided.

Oral calcium preparations include calcium _____, calcium _____, and calcium _____.	QUESTION 145
lactate . . . chloride . . . gluconate	ANSWER
Infiltration of IV solutions containing calcium may lead to tissue _____.	QUESTION 146
sloughing (necrosis)	ANSWER

QUESTION 147 Particular caution must be used with calcium and _____ medications because of the danger of _____ toxicity.

ANSWER digitalis . . . digitalis

SUMMARY OF UNIT THREE

Total electrolyte concentration of body fluids is primarily determined by "what the kidney keeps." The kidney has the ability to excrete or retain specific ions in a manner designed to maintain homeostatic conditions in the various fluid compartments. This specific function of the kidney is accomplished by diffusion, active transport, and osmosis—membrane transport mechanisms which reabsorb and secrete electrolytes and water across the tubular epithelium. Regulation of function is mediated by hormones from various endocrine glands.

As glomerular filtrate progresses through the proximal tubules, loop of Henle, distal tubule, and collecting ducts, its composition is changed. As fluid courses through the lumen of the kidney, substances are reabsorbed or secreted along the way by action of the renal epithelium. Sodium, glucose, amino acids, calcium, potassium, and phosphate are reabsorbed by active transport. These substances are transported from the tubular lumen into interstitial fluid and diffuse into capillary beds surrounding the nephrons. In this way, they are recovered from glomerular filtrate and returned to blood. When positive ions cross a membrane, negative ions follow along the electrical gradient.

Final modification of glomerular filtrate takes place in the distal convoluted tubule and collecting ducts. Water is reabsorbed under the influence of antidiuretic hormone (ADH). When ADH secretion is low, distal convoluted tubular epithelium and collecting ducts are relatively impermeable to water. Little reabsorption takes place. On the other hand, when ADH secretion is increased, tubular permeability is increased and large quantities of water are reabsorbed.

The final stages in the reabsorption of sodium take place in distal convoluted tubules and collecting ducts under the influence of aldosterone. In addition to reabsorption of sodium and water, the distal convoluted tubule and collecting duct secrete hydrogen ions, potassium ions, and ammonia, removing them from body fluids by transport from interstitial fluid into tubular lumen.

In review, the kidneys control extracellular fluid volume and concentration by
  (1) removal or retention of specific electrolytes from glomerular filtrate,
  (2) maintaining the osmolarity of body fluids at a level which is isotonic with intracellular fluids,
  (3) regulating the volume of body fluids by varying the volume of urinary output,
  (4) maintaining blood volume by controlled reabsorption of sodium and water, and
  (5) varying the pH of urine as the result of the kidney's efforts to maintain a stable acid-base balance in body fluids.

The functioning of the kidney is quantitatively controlled by hormones. Generally, when serum concentration of a specific ion is lower than normal, this acts as a stimulus, causing increased secretion of a specific hormone which acts on the kidney to increase reabsorption of that ion. For example, reduced sodium concentration in extracellular fluids triggers release of additional quantities of aldosterone. Aldosterone acts on the kidney to cause increased reabsorption of sodium, thus returning sodium levels toward normal.

Aldosterone secretion is increased in response to low sodium or high potassium concentration in extracellular fluids. Potassium ions are both reabsorbed and secreted by tubular epithelial cells, with secretion affected by aldosterone levels. As sodium ions are reabsorbed from the lumen into interstital fluid, potassium ions or hydrogen ions are secreted in the other direction to maintain electrical neutrality. When hydrogen ions are plentiful, quantitatively more of this element enters into exchange, lowering hydrogen ion concentration. When hydrogen ion concentration is low, more potassium is secreted into the flltrate.

Heart muscle is especially sensitive to an increase in potassium levels in extracellular fluids. Hyperkalemia decreases the resting membrane potential and depresses the voltage of the action potential. Contraction of heart muscle is subsequently weakened. Myocardial tissue becomes increasingly flaccid and dilates, owing to incomplete systole, or contraction. Heart muscle weakened by hyperkalemia is then subject to dangerous arrhythmias.

Parathyroid hormone is secreted in response to low calcium levels, causing reabsorption of calcium and phosphorus from bone and excretion of phosphate, thus protecting against hypocalcemia. Calcitonin, secreted in response to high serum calcium levels, inhibits demineralization of bone, thus protecting against hypercalcemia. One of the functions of calcium is regulation of the permeability of cell membranes. Hypocalcemia is accompanied by increased permeability of nerve cell membranes to sodium. Sodium leaks across membranes with greater ease, causing these membranes to depolarize easily. Continued repetitive depolarization causes tetanic contraction of innervated muscles and can lead to death by spastic respiratory paralysis.

The various electrolyte imbalances show characteristic sets of signs and symptoms which the careful observer can detect. Hypernatremia (or dehydration, or uncompensated hyperosmolarity) is followed by hypervolemia, hypertension, thirst, oliguria, dryness of the mucous membranes, and elevated temperature. Hyponatremia (or water intoxication, or uncompensated hypo-osmolarity) is reflected in muscular weakness, nausea, vomiting, and polyuria. With either extreme in sodium levels, the patient may become restless and disoriented. Coma may ensue in hypernatremia; headache and convulsions may manifest cerebral edema in hyponatremia.

Abnormalities in serum potassium levels result in malfunction of nerve and muscle tissue. A slight degree of hyperkalemia brings on gastrointestinal disturbances and diarrhea because of neuronal irritability. Severe hyperkalemia leads to flaccid paralysis following neuromuscular depression. Potentially lethal effects of hyperkalemia are the myocardial flaccidity, bradycardia, and arrhythmias accompanying impaired conduction through the Purkinje system and myocardium. Extreme hyperkalemia ends in cardiac arrest in diastole.

Hypokalemia is manifested in neuromuscular, cardiovascular, and gastrointestinal malfunction. Because hypokalemia produces an alteration in ionic ratio across nerve cell membranes, the affected patient is subject to muscular weakness, flaccid paralysis, and shallow respiration. The patient is apt to become hypotensive as vascular tone is lost, and he or she may experience gastrointestinal atony and abdominal distention. As cellular potassium is replaced with sodium and hydrogen ions, the patient tends to show acid-base disturbances. As in hyperkalemia, the most serious effect is impaired heart function: hypokalemia leads to arrhythmias, heart block, congestive heart failure, and possible cardiac arrest. Particular attention should be paid to the congestive heart failure patient on digitalis. Because digitalis is potentiated by hypokalemia, the hypokalemic digitalis patient may show signs of digitalis toxicity even though serum levels of the drug are within the non-toxic range.

Signs and symptoms of hypercalcemia appear in the neuromuscular and gastrointestinal systems as the excess calcium ions physically clog the pores of nerve and muscle cells and impair the sodium entry necessary for depolarization. The affected patient may

ELECTROLYTE BALANCE

develop nausea and vomiting, together with diarrhea or constipation, because of interference with gastrointestinal reflexes. If the hypercalcemia is due to reabsorption of calcium from bone, pathologic fractures and osteoporosis may be sustained. Renal calculi, or kidney stones, are another sign of excess calcium ions in body fluids.

A depletion of calcium ions in body fluids leads to facilitated and spontaneous depolarization of nerve and muscle tissue. Hypocalcemia is manifested in muscle spasms, leading to tetany; vascular spasms, leading to numbness and tingling; convulsions; and myocardial hyperirritability, leading to arrhythmias.

QUESTION 148 Total electrolyte concentration of body fluids is primarily determined by what is retained by the _____.

ANSWER kidney

QUESTION 149 The kidneys maintain homeostasis partially by _____ of needed electrolytes. This process transports these substances from glomerular filtrate into the _____ as well as into _____ fluid.

ANSWER reabsorption ... blood ... interstitial

QUESTION 150 The kidneys also secrete excess quantities of hydrogen ions, potassium, and other substances. This process transports substances from _____ into _____.

ANSWER blood ... glomerular filtrate

QUESTION 151 Reabsorption of water takes place in the _____ convoluted tubules and _____ ducts under the influence of _____ hormone.

ANSWER distal ... collecting ... antidiuretic

QUESTION 152 Sodium is reabsorbed in increased quantities with high circulating levels of the hormone _____.

ANSWER aldosterone

QUESTION 153 Serum calcium levels are increased in response to increased circulating levels of _____ hormone.

ANSWER parathyroid

SUMMARY OF UNIT THREE

QUESTION 154

Serum potassium levels are the result of an exchange mechanism in which either _____ ions or _____ ions are secreted as sodium is _____.

ANSWER

hydrogen ... potassium ... reabsorbed

QUESTION 155

The symptomatology of hypernatremia is similar to that of _____; the symptomatology of hyponatremia is similar to that of _____.

ANSWER

dehydration (hyperosmolarity) ... overhydration (hypo-osmolarity)

QUESTION 156

Among the signs and symptoms of uncompensated hypernatremia are increased blood _____, dry _____ membranes, _____ temperature, and restlessness and depression leading to _____.

ANSWER

viscosity ... mucous ... elevated ... coma

QUESTION 157

The hypernatremic patient manifests *hypervolemia/hypovolemia* (circle one), *hypertension/hypotension* (circle one), and *polyuria/oliguria* (circle one).

ANSWER

hypervolemia ... hypertension ... oliguria

QUESTION 158

The hyponatremic patient is *polyuric/oliguric* (circle one).

ANSWER

polyuric

QUESTION 159

Signs and symptoms of hyponatremia include muscular _____; nausea and _____; restlessness, headache, and _____ leading to _____.

ANSWER

weakness ... vomiting ... disorientation ... convulsions

QUESTION 160

A patient with hyperkalemia may develop *diarrhea/constipation* (circle one), *flaccid/spastic* (circle one) paralysis, and *tachycardia/bradycardia* (circle one).

ANSWER

diarrhea ... flaccid ... bradycardia

ELECTROLYTE BALANCE

QUESTION 161 The hypokalemic patient may develop muscular _____, _____ paralysis, and shallow _____.

ANSWER weakness ... flaccid ... respirations

QUESTION 162 Gastrointestinal _____ and abdominal _____ may characterize the hypokalemic patient.

ANSWER atony ... distention

QUESTION 163 A patient with hypokalemia may have *hypertension/hypotension* (circle one).

ANSWER hypotension

QUESTION 164 A potentially lethal effect of either hyperkalemia or hypokalemia is cardiac _____ leading to possible cardiac _____.

ANSWER arrhythmias ... arrest

QUESTION 165 The effects of digitalis are potentiated by _____.

ANSWER hypokalemia

QUESTION 166 Hypercalcemia leads to *impaired/facilitated* (circle one) depolarization, whereas hypocalcemia leads to *impaired/facilitated* (circle one) depolarization.

ANSWER impaired ... facilitated

QUESTION 167 A hypercalcemic patient may develop nausea, _____, pathologic _____, and renal _____.

ANSWER vomiting ... fractures ... calculi

Impairment of gastrointestinal reflexes in hypercalcemia may result in _____ ___ or _____.

QUESTION 168

diarrhea ... constipation

ANSWER

A hypocalcemic patient may have muscle spasms leading to _____, vascular spasms leading to _____, and myocardial hyperirritability leading to _____.

QUESTION 169

tetany ... numbness ... arrhythmias

ANSWER

UNIT FOUR • **REGULATION OF HYDROGEN ION CONCENTRATION**

HYDROGEN ION CONCENTRATION AND ACID-BASE BALANCE

ITEM 1

Unit Two dealt primarily with fluid balance, and Unit Three with electrolyte balance. The present unit examines regulatory controls for hydrogen ion concentration, or acid-base balance.

Two written symbols employed in discussing acid-base balance are $[H^+]$ and pH. The first, $[H^+]$, means simply "hydrogen ion concentration." The symbol pH is the standard abbreviation for expressing hydrogen ion concentration. In the strict sense, pH does not denote actual hydrogen ion concentration, but rather the *logarithm of the reciprocal of the hydrogen ion concentration*, or the negative log of the hydrogen ion concentration. Quantitatively, pH means the value obtained from the following formula:

$$pH = \log \frac{1}{[H^+]}, \text{ or } -\log [H^+]$$

The actual hydrogen ion concentration of arterial blood is 4×10^{-8} equivalents per liter. The pH of arterial blood is 7.4. Biochemists have adopted the pH symbol to get around the rather difficult way in which hydrogen ion concentration would otherwise have to be written. Obviously, pH 7.4 is a much simpler expression than the log $\frac{1}{4 \times 10^{-8}}$.

The symbol $[H^+]$ means _____. — **QUESTION 1**

hydrogen ion concentration — **ANSWER**

The symbol pH represents a formula used for expressing the _____ _____. — **QUESTION 2**

hydrogen ion concentration — **ANSWER**

pH is the accepted chemical symbol for quantifying hydrogen ion concentration. This symbol represents a formula, which, in turn, denotes the actual concentration of hydrogen ions. The formula represented by the symbol pH is — **QUESTION 3**

A. $pH = -\log \frac{1}{[H^+]}$

B. $pH = \log \frac{1}{[H^+]}$

The correct answer is B. The pH is the logarithm of the reciprocal of the hydrogen ion concentration. $\log \frac{1}{[H^+]}$ means the same as $-\log [H^+]$. — **ANSWER**

158 REGULATION OF HYDROGEN ION CONCENTRATION

QUESTION 4

Instead of using the actual concentration of hydrogen ions in body fluids, it is easier to think in terms of pH. The concentration of hydrogen ions in arterial blood is approximately 4×10^{-8} equivalents per liter. The pH of arterial blood is

A. $\log 4 \times 10^{-8}$

B. 7.4

ANSWER

The correct answer is B. The pH of arterial blood is the negative log of the hydrogen ion concentration, $-\log 4 \times 10^{-8}$, or the logarithm of the *reciprocal* of the hydrogen ion concentration; that is, $\log \dfrac{1}{4 \times 10^{-8}}$, or 7.4.

ITEM 2 pH, HYDROGEN IONS, AND ACIDITY

It is important to remember that *a low pH means a high concentration of hydrogen ions and that, conversely, high pH means low concentration of hydrogen ions.* Since the degree of acidity is determined by the actual quantity of free hydrogen ions in a solution, a lower pH (higher concentration of hydrogen ions) means greater acidity. On the other hand, a higher pH (lower hydrogen ion concentration) means less acidity, or greater alkalinity.

Whether or not a particular fluid is termed acid or alkaline depends on whether the pH is below 7 or above 7. At pH 7, the number of hydrogen ions is balanced by an equal number of hydroxyl (OH^-) ions, and the solution is neutral. If the pH is below 7, such as pH 2 or pH 6.5, then that solution is acid. The lower the number, the greater the acidity.

If the pH is above 7, the solution is alkaline. Another term used synonymously with "alkaline" is "basic." Thus, a fluid at pH 7.5 is an alkaline solution, or a basic solution.

QUESTION 5

The relationship of pH to acid-base balance is as follows:

A. low pH means high concentration of hydrogen ions

B. high pH means high concentration of hydrogen ions

ANSWER

The correct answer is A. As hydrogen ion concentration increases, pH decreases. High hydrogen ion concentration is indicated by low pH.

QUESTION 6

At pH 7, the number of hydrogen ions

A. is zero

B. is balanced by an equal number of hydroxyl ions

ANSWER

The correct answer is B. At pH 7, the solution contains an approximately equal number of hydrogen ions and hydroxyl ions, and is neutral.

pH 5.5 is more acid (more hydrogen ions) than

A. pH 6.0

B. pH 5.0

QUESTION 7

The correct answer is A. The number 5.5 is lower than 6.0. A solution with a lower pH (more hydrogen ions) is relatively more acid.

ANSWER

IMPORTANCE OF ACID-BASE BALANCE

ITEM 3

Disastrous alterations in body functions follow any change in the normal range of pH of body fluids. Specifically, biochemical reactions which form the basis for physiological activity are extremely sensitive to slight changes in hydrogen ion concentration.

Enzymatically catalyzed metabolic reactions, which form the basis for cell function, are pH-dependent, as we shall see later. When hydrogen ion concentration fluctuates drastically, enzyme activity slows or stops. Perhaps the most vulnerable of the organ systems is the nervous system, in which an increase in pH produces hyperexcitability and tetany and a decrease in pH causes coma.

The hydrogen ion content of body fluids needs to be regulated within very narrow limits of normal because

A. many biochemical reactions are dependent on optimum pH

B. of the lethal effects of any variation of plasma pH

QUESTION 8

The correct answer is A. Most chemical reactions in the body are enzymatically catalyzed. Since enzyme activity is extremely sensitive to changes in hydrogen ion concentration, a fluctuation in pH will accelerate some reactions and inhibit others. While it is true that drastic fluctuation of pH is lethal, it is not correct that any variation at all of plasma pH is lethal.

ANSWER

When body fluids are acidotic, the hydrogen ion concentration is abnormally

A. high

B. low

QUESTION 9

The correct answer is A. A high hydrogen ion concentration means low pH and increased acidity.

ANSWER

ITEM 4 pH IN ACIDEMIA AND ALKALEMIA

Certain mechanisms, to be explained in the next several Items, stabilize pH of body fluids within a narrow normal range. The pH of arterial blood is 7.4; that of venous blood, because of its higher carbon dioxide content, is slightly lower, at 7.35. When arterial pH drops below 7.4, the patient is in a state of *acidemia*; when arterial pH is above 7.4, the patient is in a state of *alkalemia*. A pH below 7.0 or above 7.8 is lethal. The acidotic person dies in coma; the alkalotic, in convulsions.

QUESTION 10
A pH for arterial blood below 7.4 constitutes a condition of _____; above 7.4, a condition of _____.

ANSWER
acidemia ... alkalemia

QUESTION 11
When pH for arterial blood is below 7.0, the body becomes _____ and the patient dies in _____; when above 7.8, the body becomes _____ and the patient dies in _____.

ANSWER
acidotic ... coma ... alkalotic ... convulsions

QUESTION 12
The pH of venous blood is lower than the pH of arterial blood because of

A. extra quantities of CO_2 in intracellular fluids

B. lack of oxygen in intracellular fluids

ANSWER
The correct answer is A. Remember that venous blood carries carbon dioxide from cells to the lungs. This extra carbon dioxide increases the hydrogen ion concentration and reduces pH.

ITEM 5 HOW BUFFERS STABILIZE pH

The three main lines of defense against changes in pH of body fluids are (1) buffers, (2) the respiratory system, and (3) the kidneys.

Buffers, present in all body fluids, are solutions of two or more chemical compounds which combine with strong acid or with strong base to form weak acid or weak base, thus minimizing any change in pH. Buffers combine with strong acid to restrict any decrease in pH to a minimum; they combine with strong base to restrict any increase in pH. The buffer system acts very rapidly.

List three mechanisms for control of acid-base balance:

(1) ―――――――――

(2) ―――――――――

(3) ―――――――――

QUESTION 13

---

(1) buffers

(2) respiratory system

(3) renal mechanisms (or kidney)

ANSWER

The buffer system provides a means whereby acid can be added to body fluids with a minimum _____ in pH, and alkali can be added with a minimum _____ in pH.

QUESTION 14

---

decrease ... increase

ANSWER

Buffers minimize change in pH with the addition of an acid or base by converting a _____ acid to a _____ acid, or by converting a _____ base to a _____ base.

QUESTION 15

---

stronger ... weaker ... stronger ... weaker

ANSWER

HOW THE RESPIRATORY SYSTEM STABILIZES pH

ITEM 6

The respiratory system controls hydrogen ion levels in body fluids by varying the rate of carbon dioxide removal. Carbon dioxide affects pH according to the following formula:

$$CO_2 + H_2O \xrightleftharpoons{\text{carbonic anhydrase}} H_2CO_3 \rightleftharpoons H^+ + HCO_3^-$$

Reading this reversible equation from right to left, we can see that hydrogen ions are removed from body fluids as carbon dioxide is removed by the lungs.

Increased alveolar ventilation increases carbon dioxide removal and lowers hydrogen ion concentration. Conversely, a decrease in respiratory rate and depth lessens the removal of carbon dioxide and slows removal of hydrogen ions. The respiratory mechanism, not so rapid as buffers, still works in a matter of minutes. The enzyme carbonic anhydrase, indicated in the formula above, is present in tissue cells and red blood cells. This enzyme catalyzes the reaction to make it very rapid.

QUESTION 16

The respiratory system regulates hydrogen ion concentration by varying the rate of removal of _____ _____.

ANSWER

carbon dioxide

QUESTION 17

Increased alveolar ventilation increases the rate of removal of _____ _____ and raises the _____ of body fluids.

ANSWER

carbon dioxide ... pH

QUESTION 18

Decreased alveolar ventilation decreases the rate of removal of _____ _____ and lowers the _____ of body fluids.

ANSWER

carbon dioxide ... pH

ITEM 7 HOW KIDNEYS STABILIZE pH

The kidneys regulate acid-base balance by varying the rate of removal of hydrogen ions and base bicarbonate from body fluids. As a result of the kidney's varying excretion of hydrogen ions, pH of urine becomes more or less acid. This mechanism provides a slower means of control, one that requires several hours to adjust acid-base balance. This control, however, forms a very powerful base line of activity in ridding body fluids of large quantities of either acid or base.

QUESTION 19

Renal mechanisms remove excess acid or base from body fluids by varying the quantity of _____ _____ excreted in _____.

ANSWER

hydrogen ions ... urine

QUESTION 20

The least rapid mechanism for stabilizing hydrogen ion concentration is the _____ mechanism; the most powerful is the _____ mechanism.

ANSWER

renal (kidney) ... renal (kidney)

When the kidney removes larger quantities of hydrogen ions from body fluids, this increased removal is reflected in _____ urinary pH.	QUESTION 21
decreased	ANSWER

BRÖNSTED'S DEFINITIONS OF ACID AND BASE   ITEM 8

We have defined a buffer as a combination of two or more chemical compounds which combine with either a strong acid or a strong base to form a weak acid or a weak base, thus minimizing change in pH. Brönsted's definitions of acid and base help clarify how the buffer systems work in stabilizing pH.

Brönsted defined an acid as a "proton donor" and a base as a "proton acceptor," symbolizing these definitions in the equation

$$HA \rightleftharpoons H^+ + A^-$$

where HA stands for the undissociated acid, the proton donor; H^+, the proton; and A^-, the base, or proton acceptor.

A buffer is a combination of two or more chemical compounds which A. vary the pH of body fluids by combining with acid or base B. combine with strong acid or strong base in such a way that a weak acid and weak base are produced	QUESTION 22
The correct answer is B. A buffer combination reacts with a strong acid to form a weak acid. It also reacts with a strong base to form a weak base.	ANSWER
Brönsted defined an acid as a proton _____ and a base as a proton _____.	QUESTION 23
donor . . . acceptor	ANSWER
In symbolic representation, acid is symbolized as _____; base, as _____.	QUESTION 24
HA . . . A^-	ANSWER

ITEM 9 DISSOCIATION OF ACIDS IN THE BRÖNSTED FORMULA

Acids are classified as strong or weak according to the degree of dissociation. A strong acid dissociates to a greater extent than a weak acid. In the Brönsted equation, when HA represents a strong acid, there is a relatively greater quantity of free hydrogen ions (H^+) in solution than undissociated acid (HA). The equation would be represented as $HA \rightleftharpoons H^+ + A^-$, where the long arrow to the right indicates the greater tendency of the strong acid to dissociate, releasing more hydrogen ions.

A weaker acid dissociates to a lesser degree and is relatively lower in free hydrogen ions (H^+) compared to undissociated acid (HA). A weaker acid would be represented as $HA \rightleftharpoons H^+ + A^-$, where the short arrow to the right indicates the lesser tendency of the weak acid to dissociate, releasing fewer hydrogen ions.

The degree of dissociation is the ratio of dissociated ions to undissociated molecules, and is represented as $\frac{[H^+][A^-]}{[HA]}$. Each acid has its own dissociation ratio (K) which is a known and constant value, depending on the acid and on the degree of dilution.

QUESTION 25   Acids are classified as strong or weak, depending on the extent of _____.

ANSWER   dissociation

QUESTION 26   The ratio of dissociated ions to undissociated molecules is a _____ value for each acid and for varying dilutions.

ANSWER   constant

ITEM 10 BUFFER SYSTEMS IN BODY FLUIDS

A buffer is a solution composed of an acid and the salt of that acid in a ratio such that the solution resists change in pH when a stronger acid or a stronger base is added. The principal buffers in extracellular fluids are the bicarbonate buffer system and the serum proteins. Buffers present in intracellular fluids include the phosphate, hemoglobin, and protein systems. The main buffers in urine are the bicarbonate and phosphate systems. The components of the various systems are as follows:

$$\frac{\text{sodium bicarbonate}}{\text{carbonic acid}} \quad \text{or} \quad \frac{\text{NaHCO}_3}{\text{H}_2\text{CO}_3}$$

$$\frac{\text{disodium phosphate}}{\text{monosodium phosphate}} \quad \text{or} \quad \frac{\text{Na}_2\text{HPO}_4}{\text{NaH}_2\text{PO}_4}$$

$$\frac{\text{sodium salt of oxyhemoglobin}}{\text{reduced oxyhemoglobin}} \quad \text{or} \quad \frac{\text{NaHbO}_2}{\text{HHbO}_2}$$

$$\frac{\text{sodium proteinate}}{\text{acid protein}} \quad \text{or} \quad \frac{\text{NaPr}}{\text{HPr}}$$

QUESTION 27

A buffer is a combination of an acid and a salt of that acid in a proportion which can react with strong _____ or strong _____ with very little change in the _____ of the solution.

ANSWER

acid ... base ... pH

QUESTION 28

The bicarbonate buffer system is composed of _____ and _____.

ANSWER

sodium bicarbonate ... carbonic acid

QUESTION 29

The phosphate buffer system is composed of _____ phosphate and _____ phosphate.

ANSWER

disodium ... monosodium

QUESTION 30

Other buffers in blood are _____ and _____ forms of hemoglobin and protein.

ANSWER

acid ... base

ITEM 11 — CHEMICAL REACTIONS IN THE BICARBONATE BUFFER SYSTEM

The bicarbonate buffer system is composed of a very weak acid, carbonic acid (H_2CO_3), and a salt of that acid, sodium bicarbonate ($NaHCO_3$). If a strong acid—for example, hydrochloric acid (HCl)—is added to this buffer, the hydrochloric acid reacts with the sodium bicarbonate to form an additional quantity of carbonic acid. A strong acid (HCl) is thus replaced by a weak acid (H_2CO_3). Carbonic acid dissociates to a lesser extent and, therefore, releases fewer hydrogen ions. As a result there is very little change in pH. This phase of the bicarbonate buffering action is expressed as follows:

$$NaHCO_3 + HCl \longrightarrow H_2CO_3 + NaCl$$

A strong base—for example, sodium hydroxide (NaOH)—can also be added to this buffer pair without causing any appreciable change of pH. The hydroxyl ion of the sodium hydroxide combines with the hydrogen ion of the carbonic acid to form water and sodium bicarbonate. A strong base (NaOH) has been replaced by a weak base ($NaHCO_3$), thereby preventing any significant change in pH. Expressed in equation form:

$$H_2CO_3 + NaOH \longrightarrow NaHCO_3 + H_2O$$

QUESTION 31 Addition of strong acid to the bicarbonate buffer system converts some of the base bicarbonate to _____ acid and decreases the ratio of salt to acid with little change in _____.

ANSWER carbonic ... pH

QUESTION 32 Addition of strong base to the bicarbonate buffer system converts some of the _____ acid to base bicarbonate and increases the ratio of salt to acid with little change in _____.

ANSWER carbonic ... pH

QUESTION 33 The chemical reactions occurring in the bicarbonate buffer system when strong acid is added are represented in equation form as follows: $HCl + NaHCO_3 \longrightarrow H_2CO_3 + NaCl$. Addition of the strong acid—_____ acid—to the bicarbonate buffer system converts some of the _____ to _____ acid and lowers the ratio of base to acid.

ANSWER hydrochloric ... bicarbonate ... carbonic

QUESTION 34

The chemical reactions occurring in the bicarbonate buffer system when strong base is added are represented in equation form as follows: $NaOH + H_2CO_3 \longrightarrow NaHCO_3 + H_2O$. Addition of the strong base _____ _____ to the bicarbonate buffer system converts some of the _____ acid to _____ and raises the ratio of the base to acid.

ANSWER

sodium hydroxide . . . carbonic . . . bicarbonate

RATIO OF BASE TO ACID CHANGES   ITEM 12

To illustrate the increase or decrease in concentration of base and acid in the two phases of the buffer action of the bicarbonate system (weakening a strong acid and weakening a strong base), arrows can be used. An arrow pointing upward ($\nearrow$) represents, in the equations below, an increase in the total quantity of the weak acid or weak base, as the case may be; an arrow pointing downward ($\searrow$), a decrease.

When a strong acid is added to the bicarbonate buffer pair, the strong acid reacts with the basic component. Specifically, addition of a strong acid to the bicarbonate system decreases the base component ($\searrow$) and increases the total quantity of the weak acid component ($\nearrow$):

$$HCl + \frac{NaHCO_3}{H_2CO_3} \longrightarrow \frac{NaHCO_3 \searrow}{H_2CO_3 \nearrow} + NaCl$$

Addition of a strong base causes an increase in the weak base component and a decrease in the acid:

$$NaOH + \frac{NaHCO_3}{H_2CO_3} \longrightarrow \frac{NaHCO_3 \nearrow}{H_2CO_3 \searrow} + H_2O$$

When the ratio of acid to base in any buffer system is near 1, meaning an approximately equal concentration of each of the buffer components, the buffer capacity is maximum. That is, it can react equally well with either acid or base without rapidly depleting either buffer component.

QUESTION 35

Comparison of the quantity of base in a buffer system to the quantity of acid in that system is called the base/acid _____.

ANSWER

ratio

QUESTION 36	The relative proportion, or the base/acid ratio, can be changed by an increase or decrease in either the _____ or _____ component.
ANSWER	base (numerator) ... acid (denominator)
QUESTION 37	Buffering capacity is maximum when the system can react equally well with either _____ or _____.
ANSWER	acid ... base
QUESTION 38	Buffers are most effective when the ratio of base concentration to acid concentration is _____.
ANSWER	one (equal)

ITEM 13 THE PHOSPHATE BUFFER SYSTEM

Another of the principal buffer systems of body fluids is the phosphate buffer, composed of disodium phosphate and monosodium phosphate (Na_2HPO_4/NaH_2PO_4).

Strong acid added to the phosphate buffer system converts some of the disodium phosphate to monosodium phosphate, so that, again, a weak acid replaces a strong acid with little change in pH of the solution. The reaction equation is:

$$HCl + \frac{Na_2HPO_4}{NaH_2PO_4} \longrightarrow \frac{Na_2HPO_4 \downarrow}{NaH_2PO_4 \uparrow} + NaCl$$

The quantity of Na_2HPO_4 is decreased; NaH_2PO_4, increased.

If a strong base, sodium hydroxide, is added, the sodium hydroxide reacts with the monosodium phosphate (the more acidic of the two phosphates) forming a weak base, disodium phosphate, and water. Expressed as an equation:

$$NaOH + \frac{Na_2HPO_4}{NaH_2PO_4} \longrightarrow \frac{Na_2HPO_4 \uparrow}{NaH_2PO_4 \downarrow} + H_2O$$

Since the ratio of disodium phosphate and monosodium phosphate is near 1, equal quantities of either acid or base can be added without rapid depletion of the buffering power.

Buffer systems of body fluids include the _____ buffer system, the _____ buffer system, and _____.

QUESTION 39

bicarbonate ... phosphate ... proteins

ANSWER

A strong acid added to the phosphate buffer system converts some of the _____ phosphate to _____ phosphate, replacing a _____ acid with a _____ acid.

QUESTION 40

disodium ... monosodium ... strong ... weak

ANSWER

A strong base added to the phosphate buffer system converts some of the _____ phosphate to _____ phosphate, replacing a _____ base with a _____ base.

QUESTION 41

monosodium ... disodium ... strong ... weak

ANSWER

When the ratio of base to acid in a buffer system is near 1, an equal quantity of strong _____ or strong _____ can be added with little change in _____ of the solution.

QUESTION 42

acid ... base ... pH

ANSWER

THE PROTEIN BUFFER SYSTEM

ITEM 14

Intracellular proteins and serum proteins constitute the third major buffer system. The protein buffers are both the most highly concentrated and the most powerful of the three buffer systems. Actually, nearly three-fourths of the buffering activity of body fluids is due to intracellular protein buffers.

A protein is composed of amino acids linked together. Some of the amino acids have free acidic radicals which dissociate into COO^- and H^+. Amino acids also have free basic radicals which dissociate into NH_3^+ and OH^-. The free hydroxyl radicals combine with hydrogen ions to form water, thus removing hydrogen ions from solution. Because proteins can react as either acid or base, they are termed "amphoteric."

Compared to the bicarbonate buffer system and the phosphate buffer system, the protein buffers are _____ concentrated and _____ powerful.

QUESTION 43

more ... more

ANSWER

170 REGULATION OF HYDROGEN ION CONCENTRATION

QUESTION 44 Free acidic radicals of amino acids dissociate into COO^- and _____; the free base radicals, into NH_3^+ and _____.

ANSWER $H^+ \ldots OH^-$

QUESTION 45 Free hydroxyl radicals (OH^-) remove hydrogen ions (H^+) from solution by combination to form _____.

ANSWER water

QUESTION 46 When proteins are called amphoteric, it means they can react as either _____ or _____.

ANSWER acid ... base

ITEM 15 RESPIRATORY REGULATION OF HYDROGEN ION CONCENTRATION

The body's second line of defense in regulating hydrogen ion concentration is alteration in the depth and rate of respiration. An increased concentration of carbon dioxide in body fluids shifts the pH in the direction of increased acidity. A relative decrease in carbon dioxide in body fluids shifts the pH toward increased alkalinity. By regulating carbon dioxide concentration through changes in the breathing rate, the respiratory system helps control hydrogen ion concentration.

The symbol "pCO_2" designates the pressure of carbon dioxide in body fluids. The average pCO_2 of venous blood is 45 mm. Hg, in contrast to an average 40 mm. Hg in arterial blood. The greater pressure in venous blood is due to the amount of carbon dioxide added to venous blood as a result of cellular metabolism.

QUESTION 47 The respiratory system aids in control of hydrogen ion concentration by increasing _____.

ANSWER alveolar ventilation (breathing rate)

THE RESPIRATORY CENTER AND CONTROL OF HYDROGEN ION CONCENTRATION

QUESTION 48

The average pCO_2 of arterial blood is 40 mm. Hg. The average pCO_2 of venous blood is

A. 35 mm. Hg

B. 45 mm. Hg

ANSWER

The correct answer is B. Pressure of carbon dioxide in venous blood is higher than pressure of carbon dioxide in arterial blood, because carbon dioxide, given off as a metabolic by-product from cells, goes to the lungs via venous blood.

FLUID PRESSURES AND CO₂ DIFFUSION — ITEM 16

Carbon dioxide is continuously being formed in cells as an end-product of energy metabolism. The pressure of carbon dioxide in intracellular fluids is about 46 mm. Hg. Since a gas diffuses from an area of high concentration to an area of low concentration, carbon dioxide has no trouble going from intracellular fluid (pCO_2 46 mm. Hg) into the capillaries (pCO_2 40 mm. Hg).

QUESTION 49

A pCO_2 of 46 mm. Hg in intracellular fluids and a pCO_2 of 40 mm. Hg in arterial blood means that carbon dioxide will diffuse in a direction from _____ to _____.

ANSWER

intracellular fluids ... arterial blood

THE RESPIRATORY CENTER AND CONTROL OF HYDROGEN ION CONCENTRATION — ITEM 17

When CO_2 accumulates in body fluids and increases [H^+] and acidity, the *respiratory center* in the medulla responds by regulating alveolar ventilation. The *carotid bodies* in the chemoreceptor system are also sensitive to increased pCO_2 and, to a lesser extent, to pO_2. When stimulated, the carotid bodies, located at the arch of the aorta and the bifurcation of the common carotid arteries, transmit nerve impulses to the medulla to activate the respiratory center. Thus the respiratory center is subject to a double stimulation: Increased [H^+] and increased CO_2 act as a direct stimulus, while impulses from the chemoreceptor system act as an indirect stimulus. When the respiratory center is stimulated, its rate of discharge accelerates to increase the *depth of inspiration* and, if CO_2 levels go sufficiently high, the *rate of respiration* as well. As increased alveolar ventilation removes more CO_2 from the lungs, serum CO_2 and [H^+] decrease. When normal pH is restored, excitation of the respiratory center recedes and respiration returns to normal depth and rate.

QUESTION 50 The respiratory center in the medulla responds to *a lowered/an elevated* (circle one) pH of body fluids by intensifying pulmonary ventilation.

ANSWER a lowered

QUESTION 51 Increased respiratory rate and depth lessen acidity by removing *hydrogen ions/carbon dioxide* (circle one) from the body.

ANSWER carbon dioxide

QUESTION 52 The carotid bodies in the chemoreceptor system aid in controlling [H^+] by transmitting _____ to the _____ when pCO_2 rises.

ANSWER impulses . . . respiratory center

ITEM 18 RELATION BETWEEN ALVEOLAR VENTILATION AND HYDROGEN ION CONCENTRATION

Just as elevated pCO_2 affects respiratory depth and rate, so variation in respiratory depth and rate can affect pCO_2 level. In the first instance, increased metabolic rate in cells raises pCO_2 and hydrogen ion concentration, effecting intensified respiratory depth and rate so as to remove more carbon dioxide from the lungs. In the second and reverse situation, if alveolar ventilation is decreased for pathological reasons, a smaller amount of carbon dioxide is expelled from the lungs, followed by increased pCO_2 and a rise in hydrogen ion concentration. The inability of the lungs to remove CO_2 can lessen the effectiveness of the respiratory mechanism in controlling hydrogen ion concentration to such a point that acidosis ensues.

The relationship between alveolar ventilation and pCO_2 can be approached from two directions: (1) the effect of _____ on _____. (2) the effect of _____ on _____.	QUESTION 53

(1) pCO_2 ... alveolar ventilation (respiration) (2) alveolar ventilation (respiration) ... pCO_2	ANSWER

When cellular metabolism increases, *smaller/larger* (circle one) quantities of carbon dioxide are added to body fluids.	QUESTION 54

larger	ANSWER

Any clinical abnormality which is accompanied by a decrease in pulmonary ventilation results in CO_2 retention and A. a decrease in hydrogen ion concentration B. an increase in hydrogen ion concentration	QUESTION 55

The correct answer is B. If alveolar ventilation is pathologically decreased, retention of CO_2 leads to an increase in hydrogen ion concentration in body fluids and a tendency to acidosis.	ANSWER

RENAL REGULATION OF ACID-BASE BALANCE — ITEM 19

The body's first two defenses against unbalanced acid-base concentrations in body fluids are the several buffer systems and respiratory control. The third means of regulating hydrogen ion concentration is the kidney. The kidneys regulate acid-base balance by increasing or decreasing the rates of four basic processes:
(1) tubular secretion of hydrogen ions
(2) reabsorption of sodium ions
(3) conservation of bicarbonate
(4) ammonia synthesis

The kidneys continually alternate the processes of reabsorption, secretion, and ammonia synthesis according to the body's demand for relief from either excess acid or excess base. When extracellular fluids are acidotic, the kidney excretes an acid urine to remove acid from the body; when alkalotic, the kidney conserves hydrogen ions, eliminates sodium bicarbonate, and excretes an alkaline urine.

174 REGULATION OF HYDROGEN ION CONCENTRATION

QUESTION 56
The kidneys regulate acid-base balance by reabsorption of _____ ions and secretion of _____ ions; by conservation of _____; and by synthesis of _____.

ANSWER
sodium ... hydrogen ... bicarbonate ... ammonia

QUESTION 57
When body fluids become abnormally acid, the kidney increases secretion of hydrogen ions. Excretion of an alkaline urine, on the other hand, occurs when

A. the ratio of base/acid in body fluids is low

B. body fluids contain relatively more base bicarbonate

ANSWER
The correct answer is B. The presence of large amounts of bicarbonate makes body fluids alkaline. The kidneys remove the excess base, and, as a result, an alkaine urine is excreted. When the ratio of base to acid is low, as Answer A states, the kidneys conserve base and excrete an acidic, not an alkaline, urine.

ITEM 20 MECHANISM OF SECRETION OF HYDROGEN IONS

Hydrogen ions are secreted into the tubular lumen of the nephron in an exchange with sodium. The sodium ions, reabsorbed in exchange for the hydrogen ions, diffuse from the tubular cells into extracellular fluids and combine with bicarbonate. Refer to Figure 23 as you follow the sequence of events involved in the secretion of hydrogen ions.

Beginning at the lower left of the diagram and proceeding clockwise, notice that (1) carbon dioxide passes from extracellular fluid into the interior of the tubular epithelial cells. Under the influence of the catalyst carbonic anhydrase, (2) CO_2

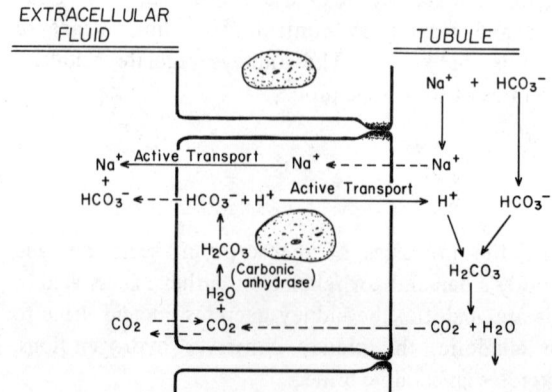

Figure 23. Chemical reactions for (1) hydrogen ion secretion, (2) sodium ion absorption in exchange for a hydrogen ion, and (3) combination of hydrogen ions with bicarbonate ions in the tubules. (From Guyton: *Textbook of Medical Physiology.* 5th Ed. Philadelphia: W. B. Saunders Company, 1976.)

combines with H_2O to form carbonic acid (H_2CO_3), which, in turn, (3) ionizes into hydrogen and bicarbonate (H^+ and HCO_3^-). The two parallel lines pointing left at the top of the diagram show (4) the reabsorbed sodium ions and bicarbonate diffusing from the tubular cells into extracellular fluid, where they combine to form sodium bicarbonate. In addition, reabsorbed sodium ions are exchanged for (5) hydrogen ions, which are secreted from epithelial cells into the lumen, as shown by the line from H^+ pointing to the right. Finally, (6) excess acid is then excreted in urine as H_2CO_3 or H_2O, as indicated by the descending lines at the lower right of the figure.

The net results of this exchange are (1) an increase in the total body $NaHCO_3$ as it is formed and conserved, and (2) a decrease in total body H_2CO_3 as it is eliminated in glomerular filtrate.

In the numbered spaces below, write the names (or symbols) of the chemicals which properly identify the corresponding numbers in the diagram below.

QUESTION 58

LABELS

1. _____   4. _____

2. _____   5. _____

3. _____   6. _____

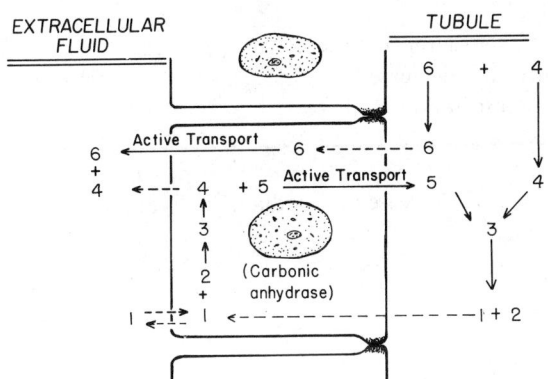

---

1. carbon dioxide (CO_2)
2. water (H_2O)
3. carbonic acid (H_2CO_3)
4. bicarbonate (HCO_3^-)
5. hydrogen (H^+)
6. sodium (Na^+)

ANSWER

The overall effect of tubular secretion of hydrogen ions is twofold:

(1) removal of _____ from body fluids

(2) retention of _____

QUESTION 59

---

hydrogen ions ... sodium bicarbonate

ANSWER

ITEM 21 ALDOSTERONE SECRETION AND REMOVAL OF HYDROGEN IONS

The ultimate quantitative control for secretion of hydrogen ions is the circulating level of aldosterone. When the adrenal glands are hyperactive, excess aldosterone is released into the blood and sodium reabsorption increases. As more sodium ions are reabsorbed, increased quantities of hydrogen ions or potassium ions are secreted through the sodium exchange mechanism. When body fluids are acidotic, hydrogen ions are usually more plentiful than potassium ions and are, therefore, preferentially secreted into glomerular filtrate and excreted in urine. A deficiency in aldosterone secretion blocks reabsorption of sodium, slowing the exchange mechanism and depressing secretion of hydrogen ions.

QUESTION 60 High circulating levels of the hormone _____ promote reabsorption of _____ and secretion of _____ ions or potassium ions.

ANSWER aldosterone ... sodium ... hydrogen

QUESTION 61 Hypoactivity of the adrenal cortex, with _____ circulating levels of aldosterone, results in decreased _____ reabsorption and decreased _____ ion secretion.

ANSWER decreased ... sodium ... hydrogen

ITEM 22 HYDROGEN AND POTASSIUM COMPETITION IN THE EXCHANGE MECHANISM

Hydrogen and potassium ions compete for exchange with sodium ions across the tubular epithelium. When body fluids are acidotic, hydrogen ions tend to be secreted in preference to potassium ions. Thus, acidosis causes a tendency toward retention of potassium, or hyperkalemia. Conversely, when extracellular fluids are hyperkalemic, and potassium ions are more prevalent than hydrogen ions, the former are secreted, causing a tendency toward hydrogen ion retention, or acidosis. A patient who is both acidotic and hyperkalemic needs therapy to combat both abnormalities.

When extracellular fluids are acidotic, relatively more _____ ions are secreted in the exchange mechanism; when extracellular fluids are hyperkalemic, relatively more _____ ions are secreted.	QUESTION 62

hydrogen . . . potassium ANSWER

Acidosis causes retention of _____ and a tendency toward _____; hyperkalemia results in retention of _____ ions and a tendency toward _____.	QUESTION 63

potassium . . . hyperkalemia . . . hydrogen . . . acidosis ANSWER

REMOVAL OF HYDROGEN IONS BY THE URINARY PHOSPHATE BUFFER

ITEM 23

In cases of extreme acidosis, hydrogen ion secretion continues as long as there is a favorable concentration gradient. One of the processes which maintain this favorable gradient is reaction with urinary buffers. The two principal urinary buffers are the bicarbonate/carbonic acid system and the disodium phosphate/monosodium phosphate system. Since the manner in which the two systems function is virtually the same, this Item will describe only the latter.

Examine Figure 24 as you study this Item. The process depicted within the epithelial cell is identical to that described in Item 20. Reread Item 20 if you need to refresh yourself on this part of the process. Directing your attention now to the top right of the diagram, notice the disodium phosphate, Na_2HPO_4, which is normally present in glomerular filtrate in a much greater quantity than monosodium phosphate, NaH_2PO_4.

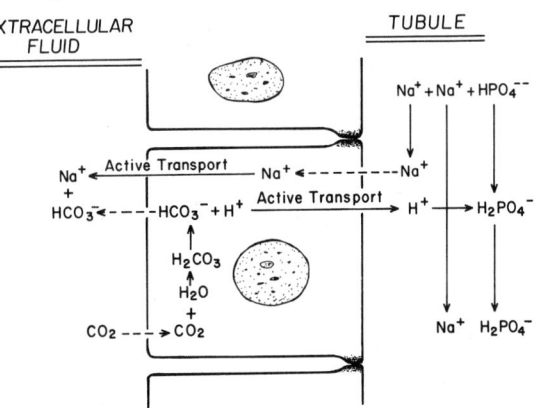

Figure 24. Chemical reactions in the tubules involving hydrogen ions, sodium ions, and the phosphate buffer system. (From Guyton: *Textbook of Medical Physiology.* 5th Ed. Philadelphia: W. B. Saunders Company, 1976.)

178 REGULATION OF HYDROGEN ION CONCENTRATION

Hydrogen ions are exchanged for the sodium ions in disodium phosphate to form monosodium phosphate, which then passes into the urine, as shown in the middle and right of the illustration. The availability of the phosphate buffer in glomerular filtrate with which H⁺ combines provides the favorable concentration gradient for secretion of hydrogen.

QUESTION 64

The principal urinary buffers are

(1) _____ / _____

(2) _____ / _____

ANSWER

(1) bicarbonate/carbonic acid

(2) disodium phosphate/monosodium phosphate

QUESTION 65

Hydrogen ions react with the phosphate urinary buffer to convert _____ phosphate to _____ phosphate.

ANSWER

disodium ... monosodium

QUESTION 66

Sodium ions liberated from the phosphate buffer cross the tubular epithelium in exchange for _____ ions.

ANSWER

hydrogen

QUESTION 67

Sodium ions combine with _____ ions and are reabsorbed into _____ fluid.

ANSWER

bicarbonate ... extracellular

QUESTION 68

In the numbered spaces below, write the names (or symbols) of the chemicals which properly identify the corresponding numbers in the diagram below. Refer to the text of the Item to complete the labeling.

LABELS

1. _____   5. _____
2. _____   6. _____
3. _____   7. _____
4. _____   8. _____

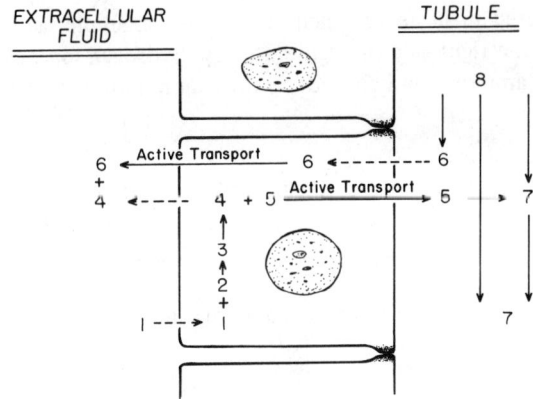

1. carbon dioxide (CO_2)
2. water (H_2O)
3. carbonic acid (H_2CO_3)
4. bicarbonate (HCO_3^-)
5. hydrogen (H^+)
6. sodium (Na^+)
7. monosodium phosphate (NaH_2PO_4)
8. disodium phosphate (Na_2HPO_4)

ANSWER

REMOVAL OF HYDROGEN IONS BY COMBINATION WITH AMMONIA

ITEM 24

In addition to reaction with urinary buffers, a second process which provides a favorable concentration gradient for secretion of excess hydrogen ions is the ammonia-combining mechanism. Ammonia (NH_3), derived from glutamic acid and other amino acids, as shown in Figure 25, is synthesized by cells of the kidney tubules and collecting ducts. It diffuses into the tubular lumen, where it combines with hydrogen ions to form ammonium ions, NH_4^+. The ammonium ions then combine with chloride (or some other

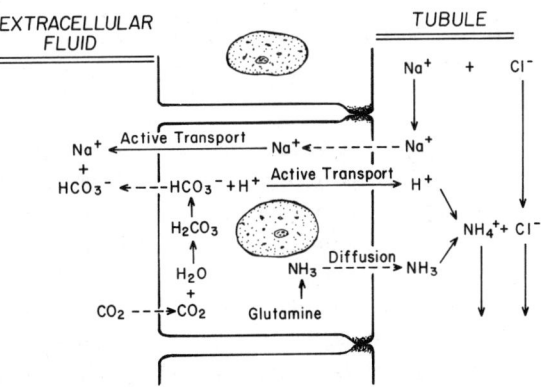

Figure 25. Secretion of ammonia by the tubular epithelial cells, and reaction of the ammonia with hydrogen ions in the tubules. (From Guyton: *Textbook of Medical Physiology.* 5th Ed. Philadelphia: W. B. Saunders Company, 1976.)

180 REGULATION OF HYDROGEN ION CONCENTRATION

anion) to be excreted in urine as a neutral salt—for example, as ammonium chloride, NH_4Cl. Notice on the right side of Figure 25 how hydrogen ions, exchanged for sodium ions, combine with ammonia and chloride to form ammonium chloride, a neutral salt.

QUESTION 69 Ammonia diffuses from renal epithelial cells into the tubular _____, combining with hydrogen ions to form _____ ions (NH_4^+).

ANSWER lumen ... ammonium

QUESTION 70 The ammonia-combining mechanism causes no appreciable fall in urinary pH because the final product, _____ _____, is a neutral salt.

ANSWER ammonium chloride (NH_4Cl)

QUESTION 71 The net effect of the ammonia-combining mechanism is
(1) removal of _____ _____
(2) retention of _____

ANSWER hydrogen ions ... sodium

QUESTION 72 Removal of hydrogen ions by reaction with urinary buffers or by combining with ammonia permits continued diffusion of _____ _____ down a _____ gradient.

ANSWER hydrogen ions ... concentration

ITEM 25 RENAL AND RESPIRATORY MUTUAL COMPENSATION

When a patient suffers some respiratory insufficiency, the renal mechanisms we have described compensate to maintain acid-base balance. The kidneys counteract respiratory acidosis by increasing the secretion of hydrogen ions, retention of bicarbonate, and synthesis of ammonia. Renal mechanisms compensate for respiratory alkalosis by repressing these processes for removing acid from the body. In compensating for respiratory acidosis, the kidneys excrete a more acid urine; in compensating for respiratory alkalosis, the kidneys excrete a more alkaline urine.

The compensation is reversed when kidney function is impaired. Alveolar ventilation varies as the body attempts to remove or retain hydrogen ions. Hyperventilation will at least partially compensate for metabolic acidosis. The breathing, for instance, of a patient in diabetic coma grows rapid as the respiratory mechanism attempts to remove the accumulation of H^+ by expiring larger volumes of carbon dioxide.

When respiratory acidosis occurs, the kidneys compensate by increasing excretion of _____. **QUESTION 73**

---

hydrogen ions **ANSWER**

When kidneys compensate for respiratory acidosis, urine becomes more _____; there is a conservation of _____; and synthesis of _____ is increased. **QUESTION 74**

---

acidic ... bicarbonate ... ammonia **ANSWER**

When respiratory alkalosis occurs, the kidneys compensate by decreasing secretion of _____ ions, excreting _____ into urine, and decreasing synthesis of _____. **QUESTION 75**

---

hydrogen ... bicarbonate ... ammonia **ANSWER**

When metabolic acidosis occurs, the respiratory system is able to compensate partially by increasing _____. **QUESTION 76**

---

alveolar ventilation **ANSWER**

Metabolic alkalosis encounters respiratory resistance in the form of *increased/ decreased* (circle one) alveolar ventilation. **QUESTION 77**

---

decreased **ANSWER**

CLINICAL ASPECTS OF ACID-BASE BALANCE

ITEM 26

Up to this point, Unit Four has examined the mechanisms regulating acid-base balance under normal body conditions. The remainder of this Unit will consider the symptoms, etiology, and treatment of pathological situations involving acid-base imbalance.

Clinical cases of acid-base imbalance fall under four headings: (1) respiratory acidosis, (2) respiratory alkalosis, (3) metabolic acidosis, and (4) metabolic alkalosis. The first abnormality, *respiratory acidosis*, results when any condition decreases effective pulmonary ventilation to such an extent that CO_2 accumulates in extracellular fluids and leads to retention of excessive amounts of hydrogen ions.

Respiratory alkalosis, which occurs much less frequently than respiratory acidosis, may result from hyperventilation or other causes that bring about loss of CO_2 from extracellular fluids at an excessively rapid rate. As $[H^+]$ decreases and pH rises, a reduction of carbonic acid raises the ratio of bicarbonate to carbonic acid, which is pathognomonic of alkalosis.

The terms *metabolic acidosis* and *metabolic alkalosis* refer to acid-base imbalances traceable to any cause other than respiratory pathology.

QUESTION 78

Place a check mark before the category of acid-base imbalance which is encountered least frequently clinically.

( ) 1. respiratory acidosis   ( ) 3. metabolic acidosis

( ) 2. respiratory alkalosis   ( ) 4. metabolic alkalosis

ANSWER

You should have placed the check in front of No. 2, respiratory alkalosis.

QUESTION 79

Respiratory acidosis is a condition in which the hydrogen ion concentration of body fluids is increased as a result of

A. respiratory retention of CO_2

B. increased formation of CO_2

ANSWER

The correct answer is A. Acidosis due to any condition which alters the respiratory system's ability to vary rate and depth of breathing in response to $[H^+]$ of body fluids is called respiratory acidosis. Increased formation of CO_2, mentioned in choice B, is a metabolic, not a respiratory, cause.

QUESTION 80

Respiratory alkalosis is a condition in which $[H^+]$ of body fluids is decreased because of

A. decreased formation of CO_2

B. excessive carbon dioxide "blow off"

ANSWER

The correct answer is B. When hyperventilation "blows off" excessive amounts of CO_2 into the air to such a degree that excessive quantities of hydrogen ions are removed and pH rises, the resulting alkalosis is of respiratory origin. Decreased formation of CO_2, on the other hand, is a metabolic, not a respiratory, cause of alkalosis.

QUESTION 81

Acidosis or alkalosis originating from any cause other than respiratory pathology is properly referred to as _____ in origin.

ANSWER

metabolic

EFFECTS OF ACID-BASE IMBALANCE ON THE CENTRAL NERVOUS SYSTEM

ITEM 27

Abnormalities in hydrogen ion concentration affect primarily the central nervous system. Acidosis leads to CNS depression, with decreased mentation, coma, and possible death. Alkalosis leads to CNS hyperexcitability, with nervousness, muscular twitchings, possible convulsions, and death. A threatened imbalance can be averted by close monitoring of pH, pCO_2, and total CO_2 content. Abnormalities in these values are indicators of whether the underlying cause is respiratory or metabolic and also of the state of renal or respiratory compensation.

Acidosis *excites/depresses* (circle one) the central nervous system, sometimes to the point of causing *convulsions/coma* (circle one), whereas alkalosis *excites/depresses* (circle one) the CNS, sometimes to the point of causing *convulsions/coma* (circle one).

QUESTION 82

---

depresses ... coma ... excites ... convulsions

ANSWER

Three important laboratory values which must be observed when acidosis or alkalosis is impending are (1) _____, (2) _____, and (3) _____.

QUESTION 83

---

pH ... pCO_2 ... CO_2

ANSWER

CLINICAL MEASUREMENT OF ARTERIAL pH

ITEM 28

Clinical practice employs various means to determine acid-base balance. Among the more critical determinations needed for diagnostic purposes are (1) pH of arterial blood, (2) pCO_2 in arterial blood, and (3) total serum CO_2 concentration.

Arterial pH is established by measuring the pH of blood with a glass electrode pH meter. Measurements must be done rapidly and with extreme caution lest dissolved carbon dioxide diffuse out of the sample and give a false alkaline reading. Normally the pH of arterial blood is 7.4.

Although pH measurement can establish the presence of either alkalosis or acidosis, it cannot tell whether the cause is respiratory or metabolic. The method of ascertaining the origin of alkalosis or acidosis will be explained in the next several Items dealing with use of the nomogram.

184 REGULATION OF HYDROGEN ION CONCENTRATION

QUESTION 84 — Arterial pH can be carefully measured with a glass electrode _____ _____.

ANSWER — pH meter

QUESTION 85 — Clinical tests used to study acid-base balance include measurements of _____ of arterial blood; of _____ in arterial blood; and of total serum _____ concentration.

ANSWER — pH ... pCO_2 ... CO_2

QUESTION 86 — In measuring arterial pH, if care is not taken to prevent diffusion of _____ _____ out of the plasma, the _____ component of the base/acid ratio could be altered and give a false _____ reading.

ANSWER — carbon dioxide ... acid ... alkaline

QUESTION 87 — Accurate measure of arterial pH reveals whether acidosis or alkalosis is present in a patient but does not tell whether the cause is _____ or _____.

ANSWER — respiratory ... metabolic

ITEM 29 USE OF THE NOMOGRAM

The origin of an acid-base imbalance is determined as respiratory or metabolic by the use of a nomogram. The nomogram is a chart which relates pH, pCO_2, and total CO_2 content. The nomogram graphically categorizes a particular combination of values as indicative of respiratory acidosis, respiratory alkalosis, metabolic acidosis, metabolic alkalosis, or some combination of these abnormalities.

Examine the nomogram depicted in Figure 26. The horizontal scale at the bottom represents pH, with normal pH range for both arterial blood (7.35 to 7.45) and venous blood (7.32 to 7.42) indicated in brackets. The horizontal scale at the top represents pCO_2, with normal pCO_2 range for both arterial blood (35 to 45) and venous blood (42 to 55) indicated at the upper right. The vertical scale at the left represents total CO_2 content in millimols per liter (mM./L.).

Curving lines slanting diagonally upward to the right represent pCO_2 in mm. Hg. The curving line with the value 40 represents the average level. Values to the left of this curving line represent increased CO_2 level; to the right, decreased CO_2 level.

Lines slanting diagonally downward from left to right represent values for total CO_2

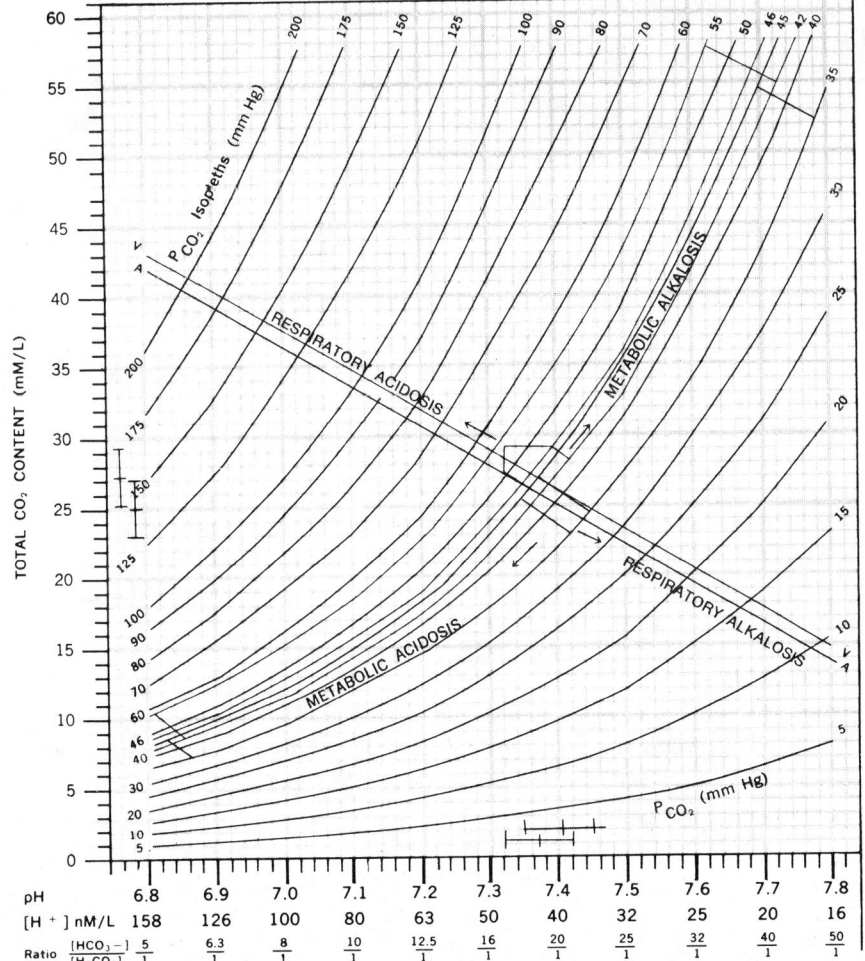

Figure 26. A nomogram (courtesy of Methodist Hospital, Memphis, Tennessee).

content, principally bicarbonate. Brackets to the left represent the normal range for arterial and venous blood.

If respiratory acidosis is present, pCO_2 is elevated; if respiratory alkalosis is present, pCO_2 is depressed. If metabolic acidosis is present, total CO_2 content is low; if metabolic alkalosis is present, total CO_2 content is high.

The normal range of pH for arterial blood (7.35 to 7.45) is slightly *higher/lower* (circle one) than the normal range for venous blood.

QUESTION 88

higher

ANSWER

186 REGULATION OF HYDROGEN ION CONCENTRATION

QUESTION 89
The normal range of pCO_2 for venous blood (42 to 55 mm. Hg) is somewhat *higher/lower* (circle one) than the normal range for arterial blood.

ANSWER
higher

QUESTION 90
On the nomogram, values to the left of the curving line with the value 40 represent increased *acidity/alkalinity* (circle one) caused by *increased/decreased* (circle one) total CO_2 content; values to the right represent increased *acidity/alkalinity* (circle one) caused by *increased/decreased* (circle one) CO_2 levels.

ANSWER
acidity ... increased ... alkalinity ... decreased

QUESTION 91
Elevated pCO_2 is indicative of respiratory *acidosis/alkalosis* (circle one); low pCO_2 is indicative of respiratory *acidosis/alkalosis* (circle one).

ANSWER
acidosis ... alkalosis

QUESTION 92
High total CO_2 content is indicative of metabolic *acidosis/alkalosis* (circle one); low total CO_2 content is indicative of metabolic *acidosis/alkalosis* (circle one).

ANSWER
alkalosis ... acidosis

QUESTION 93
The level of pCO_2 is the value observed in checking for *respiratory/metabolic* (circle one) involvement in acid-base imbalance; total CO_2 content is the significant value in checking for *respiratory/metabolic* (circle one) involvement.

ANSWER
respiratory ... metabolic

ITEM 30 ILLUSTRATIONS OF THE USE OF THE NOMOGRAM

Look back to Figure 26 and notice on the nomogram that if pCO_2 is elevated and total CO_2 content is normal or low, pH will be low. Such a combination of values indicates uncompensated respiratory acidosis or the simultaneous presence of both respiratory and metabolic acidosis.

Suppose pCO_2 is high and total CO_2 content is above normal. If arterial pH is within the normal range, there is an indication here of either compensated respiratory acidosis or a combination of respiratory acidosis and metabolic alkalosis.

If total CO_2 content is elevated and pCO_2 is normal, pH will be in the alkalotic range. This combination of values indicates either uncompensated metabolic alkalosis or the simultaneous presence of both metabolic and respiratory alkalosis.

If total CO_2 is high and pCO_2 is correspondingly high, then pH will approach normal, indicating compensated metabolic alkalosis or a combination of metabolic alkalosis and respiratory acidosis.

Refer to the nomogram in Figure 26 to answer the questions on this Item.

Elevated pCO_2 and depressed total CO_2 content result in *an increased/a decreased* (circle one) pH.	QUESTION 94
a decreased	ANSWER
Uncompensated respiratory alkalosis is characterized by _____ pCO_2, _____ total CO_2 content, and _____ pH.	QUESTION 95
decreased ... normal ... increased	ANSWER
Elevated total CO_2 content and normal pCO_2 result in *an increased/a decreased* (circle one) pH.	QUESTION 96
an increased	ANSWER
Uncompensated metabolic acidosis is characterized by _____ total CO_2 content, _____ pCO_2, and _____ pH.	QUESTION 97
decreased ... normal ... lowered	ANSWER

CASE STUDY—CHECKING ACID-BASE BALANCE ON THE NOMOGRAM

ITEM 31

Patient A has an arterial pH of 7.40. On the nomogram in Figure 26 locate the scale with pH 7.40 and draw a line upward. Patient A's pCO_2 in arterial blood is 40 mm. Hg. Locate the 40 position in the upper right corner and follow the sloping line downward and to the left. Place a dot at the point where the vertical line from pH 7.40 and the sloping line cross. Note that the dot falls in the center of the rectangle representing normal. Now draw a horizontal line from the dot to the scale representing total CO_2 content. The value for total CO_2 content is found to be 25 mM./L., which is in the center of the bracket representing normal values for arterial blood. A check of the pertinent values for Patient A on the nomogram indicates that he has no acid-base problem.

QUESTION 98   Work out the answer to this question by plotting values on the nomogram in Figure 26. A patient has an arterial pH of 7.46 and a pCO_2 of 40. His total CO_2 content is _____ mM./L., and his acid-base abnormality is _____ _____.

ANSWER   29 ... metabolic alkalosis

ITEM 32   CASE STUDY—CHECKING ACID-BASE BALANCE IN A DIABETIC PATIENT

This somewhat longer item is presented in three sections: (1) Description of the Case, (2) Plotting Values on the Nomogram, and (3) Analysis of Findings from the Nomogram.

Description

Mr. B., a diabetic, has just entered the hospital with pneumonia. Possible acid-base complications must be checked carefully, inasmuch as diabetes is a frequent cause of metabolic acidosis. Deprived of insulin as a means of utilizing carbohydrates, the diabetic depends in part on metabolism of fat for energy. Excessive fat metabolism generates large quantities of acidic ketone bodies, which add to the total body acid and thereby contribute to acidosis. Furthermore, excretion of ketone bodies as sodium salts of the acids depletes sodium bicarbonate, further disposing the body to acidosis.

Prior to his hospitalization, Mr. B.'s diabetes was controlled by Orinase, an oral hypoglycemic supplied in 0.5-gm. tablets. Mr. B.'s dosage was 4 tablets a day, and he was on a 2000-calorie diabetic diet. On admittance, his blood pressure was 146/84, temperature 101.4, pulse 98; respirations were deep and labored and their rate was 32/min. A Clinitest done on a urine sample showed 4+ glucosuria. The patient appeared extremely lethargic and coughed frequently. Laboratory findings showed a blood glucose value of 260 mg.%, a hyperglycemic level. Arterial pH was 7.28. The pCO_2 level was 48, indicating hypercapnia.

Plotting Values on the Nomogram

On the nomogram in Figure 26, locate the scale with pH 7.28 and draw a line upward. Now locate the position of pCO_2 48 and mark its intersection with the vertical line from

CASE STUDY—CHECKING ACID-BASE BALANCE IN A DIABETIC PATIENT 189

pH 7.28. This point corresponds with 24 on the left-hand scale, indicating that the total CO_2 content is 24, which is within the range of normal.

Analysis

Mr. B. probably had an initial metabolic acidosis when the stress of the infection from pneumonia required a larger insulin dosage than he had been taking. As the pneumonia obstructed sections of the alveolar membrane, he was unable to exhale enough carbon dioxide, with the result that pCO_2 became elevated. At this time, he has a combined metabolic and respiratory acidosis. Since total CO_2 content, primarily bicarbonate, is within the normal range, the kidneys have partly compensated for the respiratory insufficiency by retaining bicarbonate and excreting an acid urine. Notice at the bottom of the nomogram that the ratio of HCO_3^- to H_2CO_3 in Mr. B.'s case is nearly 16 to 1. This ratio would indicate that, in spite of the increased acid component, the kidneys are able to retain enough bicarbonate to partly compensate.

Refer to the nomogram in Figure 26 if necessary to answer this question. An arterial pH of 7.28 is *acidotic/normal/alkalotic* (circle one), while a pCO_2 reading of 48 indicates *increased/decreased* (circle one) carbon dioxide levels.

QUESTION 99

---

acidotic . . . increased

ANSWER

Diabetes contributes to metabolic acidosis because conversion of fat to energy produces a large quantity of _____ bodies, which are chemically acid, and because sodium bicarbonate is lost in the excretion of _____ bodies.

QUESTION 100

---

ketone . . . ketone

ANSWER

Mr. B.'s initial metabolic acidosis was compounded by _____ _____, which was the result of retention of _____ due to the pulmonary insufficiency accompanying his pneumonia.

QUESTION 101

---

respiratory acidosis . . . CO_2 (carbon dioxide)

ANSWER

Mr. B.'s kidneys compensated for his acidosis by excretion of _____ _____ and retention of _____.

QUESTION 102

---

hydrogen ions . . . bicarbonate

ANSWER

ITEM 33 — SIGNS AND SYMPTOMS OF RESPIRATORY ACIDOSIS

The signs and symptoms of respiratory acidosis are manifested in cardiac and pulmonary changes. The patient may be dyspneic and cyanotic because oxygen and carbon dioxide do not diffuse easily across the respiratory membrane. The pulse rate accelerates because of the cardiovascular effects of hypercapnia, mediated by the chemoreceptor system.

QUESTION 103 The principal organs affected by respiratory acidosis are the _____ and _____.

ANSWER heart ... lungs

QUESTION 104 The cardiovascular response to hypercapnia is increased _____.

ANSWER pulse rate

QUESTION 105 Because of inadequate respiratory exchange, the patient with respiratory acidosis may be _____ and _____.

ANSWER dyspneic ... cyanotic

ITEM 34 — CAUSES OF RESPIRATORY ACIDOSIS

Respiratory acidosis is a condition in which the [H^+] of body fluids is elevated because of failure of the lungs to remove a sufficient amount of CO_2 from the body. Common causes of this condition are emphysema, bronchiectasis, pneumonia, asthma, pulmonary edema accompanying congestive heart failure, and any other condition which impairs diffusion of CO_2 across the alveolar membrane. Respiratory acidosis can also be induced by an overdose of drugs which depress the respiratory center in the medulla, such as Demerol and morphine.

In any of the pathological states which bring on respiratory acidosis, serum pCO_2 is elevated, resulting in an increase in H_2CO_3, which then dissociates to release H^+ and HCO_3^-. As some of the bicarbonate diffuses into cells, an excess of hydrogen ions is left in extracellular fluid, tending to lower pH. Actually, the extent to which pH may be lowered will be conditioned by the degree of buffer or renal compensation for the respiratory insufficiency.

Hydrogen ion concentration is *increased/decreased* (circle one) in respiratory acidosis, while pH may or may not be *increased/decreased* (circle one).	QUESTION 106
increased ... decreased	ANSWER
In any pulmonary dysfunction which brings on respiratory acidosis, serum pCO_2 becomes *elevated/depressed* (circle one), resulting in *an increase/a decrease* (circle one) in H_2CO_3.	QUESTION 107
elevated ... an increase	ANSWER
Four respiratory diseases which invite respiratory acidosis include (1) _____, (2) _____, (3) _____, and (4) _____.	QUESTION 108
emphysema ... bronchiectasis ... pneumonia ... asthma (depression of respiratory center, pulmonary edema)	ANSWER
Two drugs which depress the respiratory center are _____ and _____.	QUESTION 109
Demerol ... morphine	ANSWER
In respiratory acidosis, hydrogen ion concentration is elevated, owing to retention of _____.	QUESTION 110
carbon dioxide (CO_2)	ANSWER

COMPENSATION FOR RESPIRATORY ACIDOSIS

ITEM 35

To maintain a safe arterial pH of 7.4, the ratio of bicarbonate to carbonic acid in the buffer system must be held at 20:1, as a glance at the bottom of the nomogram in Figure 26 will show. As respiratory acidosis becomes increasingly severe, the bicarbonate component of the buffer pair continues to diminish. One means the body employs to replenish HCO_3^- is to shift extracellular chloride into intracellular fluids in exchange for intracellular bicarbonate. Thus some compensation is effected by borrowing from the intracellular pool of HCO_3^- in return for extracellular Cl^-.

Renal compensation contributes to re-establishing the 20:1 ratio of base to acid by retaining bicarbonate and excreting hydrogen ions. The kidneys accelerate ammonia secretion, thereby encouraging the removal of excess H^+ in the urine in the form of ammonium chloride.

QUESTION 111

Retention of CO_2 in respiratory acidosis results in an increase in the _____ acid component of the bicarbonate buffer system.

ANSWER

carbonic

QUESTION 112

The ratio of base to acid in the buffer system consistent with maintaining an arterial pH of 7.4 is _____ to 1.

ANSWER

20

QUESTION 113

The chloride shift involves exchange of extracellular _____ for intracellular _____ in order to raise serum _____ levels.

ANSWER

chloride ... bicarbonate ... bicarbonate

QUESTION 114

Besides retaining bicarbonate, the kidneys compensate for respiratory acidosis by increasing the secretion of _____, which combines with _____ ions to form ammonium ions that are excreted in urine as _____.

ANSWER

ammonia ... hydrogen ... ammonium chloride

ITEM 36   LABORATORY VALUES IN COMPENSATED RESPIRATORY ACIDOSIS

In compensated respiratory acidosis, serum pH is 7.4 or very slightly below; total CO_2 content, chiefly bicarbonate, is elevated above the normal of 23 to 27 mM./L.; and urine is acidic. If respiratory acidosis remains decompensated in spite of the body's stabilizing efforts, pH will steadily drop, notwithstanding the elevated bicarbonate and acid urine.

When body mechanisms successfully compensate for respiratory acidosis, total CO_2 content is *increased/decreased/normal* (circle one), pH is *increased/decreased/normal* (circle one), and urine is *alkaline/acid/normal* (circle one).

QUESTION 115

increased ... normal ... acid

ANSWER

When the compensating mechanisms fail to correct respiratory acidosis, bicarbonate is _____, but pH continues to _____.

QUESTION 116

elevated ... drop

ANSWER

TREATMENT OF RESPIRATORY ACIDOSIS

ITEM 37

Treatment of respiratory acidosis must be directed toward the pulmonary dysfunction which brings on the acidosis in the first place. If infection is the cause, antibiotics are an effective means of treatment. In emphysema or chronic obstructive pulmonary disease, supportive therapy such as oxygen in low doses, intermittent positive pressure breathing, and bronchodilators may be called for.

Certain clinical procedures may be instituted to supplement the body's compensatory mechanisms for correcting the acidotic state. If the degree of acidosis is severe, sodium bicarbonate, sodium lactate, or Ringer's lactate solution may be administered. Gastric suction may be employed to remove hydrochloric acid from the stomach, and thus further reduce total body acid.

In cases of respiratory acidosis caused by severe pulmonary disease, treatment includes such supportive therapy as administration of _____ and _____ breathing.

QUESTION 117

oxygen ... positive pressure

ANSWER

Clinical measures for correcting the acidosis itself include administration of sodium _____, sodium _____, or _____ lactate solution.

QUESTION 118

bicarbonate ... lactate ... Ringer's

ANSWER

ITEM 38 — MANAGEMENT OF HYPERKALEMIA IN RESPIRATORY ACIDOSIS

Two complications which may be encountered in respiratory acidosis are hyperkalemia and hypocalcemia. Ordinarily, potassium and hydrogen ions compete for secretory sites in the sodium exchange mechanism, but a build-up of hydrogen ions, as in respiratory acidosis, blocks the secretory machinery for excretion of potassium ions. If frank hyperkalemia develops, treatment follows this course: potassium intake is reduced or eliminated; excretion is promoted by use of diuretics such as Lasix; and removal is facilitated by use of ion exchange resins.

QUESTION 119 In the patient with respiratory acidosis, dangerous hyperkalemia can be managed by eliminating his _____ intake and by use of _____ and _____ resins.

ANSWER potassium ... diuretics ... ion exchange

ITEM 39 — MANAGEMENT OF HYPOCALCEMIA IN RESPIRATORY ACIDOSIS

A hazard encountered in the correction of respiratory acidosis is the development of hypocalcemia. Since calcium ionizes best in an acid pH and is driven into bone in alkalosis, caution must be exercised in reversing acidosis intravenously with an alkalinizing solution. If serum calcium levels drop precipitously, hypocalcemia and tetany may ensue. For this reason, calcium chloride or calcium gluconate may be given intravenously before the alkalinizing solution. Calcium compounds have a tendency to precipitate, and they should not be added to the regular IV solutions.

QUESTION 120 The reversal of acidosis by IV infusion of sodium bicarbonate poses the danger of hypocalcemia, leading to _____.

ANSWER tetany

QUESTION 121 To offset the danger of hypocalcemia in reversing acidosis through IV infusion of sodium bicarbonate, _____ _____ may be given.

ANSWER calcium chloride

SIGNS AND SYMPTOMS OF RESPIRATORY ALKALOSIS

ITEM 40

Respiratory alkalosis is accompanied by an increase in neuromuscular irritability. In certain susceptible individuals, hyperreflexia and twitching may lead to convulsions. Laboratory values show pCO_2 in arterial blood below the 35 to 45 mm. Hg normal range and an alkaline urine. Serum electrolytes tend toward hypokalemic levels, sometimes falling below the 3.8 to 4.5 mEq./L. norm. Treatment for respiratory alkalosis is aimed at eliminating the root condition causing the imbalance.

In respiratory alkalosis, the nervous system and muscle tissue become _____.

QUESTION 122

irritable

ANSWER

Signs of respiratory alkalosis include _____, _____, and possible _____.

QUESTION 123

hyperreflexia ... twitching ... convulsions

ANSWER

Laboratory values for the patient with respiratory alkalosis include:

(1) _____ pCO_2
(2) _____ urine
(3) _____ serum potassium levels

QUESTION 124

decreased ... alkaline ... low

ANSWER

CAUSES OF RESPIRATORY ALKALOSIS

ITEM 41

Respiratory alkalosis results when hyperventilation removes too much carbon dioxide and H^+ from body fluids. As CO_2 and water are lost in expired air, more and more H^+ and bicarbonate ions combine to form carbonic acid, which is dissociated and "blown off" with accelerated alveolar ventilation. The relatively larger amount of bicarbonate remaining elevates the ratio of base to acid in the bicarbonate buffer system, creating a tendency to alkalosis.

196 REGULATION OF HYDROGEN ION CONCENTRATION

Hyperventilation most often results from excitement and anxiety states. Under such excitation, the usual medullary control of CO_2 and H^+ appears to be temporarily overridden. Fever, in some instances, may so overstimulate the CNS that it induces respiratory alkalosis. Respiratory alkalosis may also result from the action of certain drugs—for example, from aspirin poisoning.

QUESTION 125

Place a check mark in front of the correct answer. Respiratory alkalosis results from

( ) 1. increased H_2CO_3

( ) 2. decreased H_2CO_3

( ) 3. increased HCO_3^-

( ) 4. decreased HCO_3^-

ANSWER

You should have checked No. 2, decrease of carbonic acid.

QUESTION 126

Fever may create a tendency to respiratory alkalosis through excessive stimulation of the _____.

ANSWER

central nervous system

QUESTION 127

Typical of the type of drug action which could lead to respiratory alkalosis is _____ poisoning.

ANSWER

aspirin

QUESTION 128

When hyperventilation follows upon anxiety, the apparent reason is a suspension of control by the _____ over _____ and hydrogen ions.

ANSWER

medulla ... carbon dioxide (CO_2)

ITEM 42   COMPENSATION FOR RESPIRATORY ALKALOSIS

Compensation for respiratory alkalosis is exactly opposite that for respiratory acidosis. Since the acid component of the bicarbonate buffer system is depleted, the kidneys retain hydrogen ions and excrete sodium bicarbonate, thus reducing the base component correspondingly. In this manner, the kidneys attempt to maintain the ratio of base to acid. In respiratory acidosis, the kidneys increase ammonia synthesis; in respiratory alkalosis, it is necessary that they reduce ammonia synthesis. With less ammonia available for combination with hydrogen ions to form ammonium, obviously there is a smaller amount of ammonium chloride excreted in urine and a larger quantity of hydrogen ions conserved.

A particular complication which may attend the compensation for respiratory alkalosis is hypokalemia. As the kidneys attempt to offset the alkalosis by conserving H^+, more K^+ is apt to be secreted in the sodium exchange, with the possible risk of total potassium being lowered to the point of hypokalemia.

Renal compensation for respiratory alkalosis involves *retention/excretion* (circle one) of H^+, *retention/excretion* (circle one) of HCO_3^-, and *an increase/a decrease* (circle one) in ammonia synthesis.

QUESTION 129

retention ... excretion ... a decrease

ANSWER

A complicating factor which may occur in the process of renal compensation for respiratory alkalosis is _____.

QUESTION 130

hypokalemia

ANSWER

SIGNS AND SYMPTOMS OF METABOLIC ACIDOSIS

ITEM 43

The effects of metabolic acidosis are seen primarily in depression of nerve function. The patient becomes apathetic, disoriented, stuporous, and comatose. Respirations are deep and rapid as the lungs attempt to compensate.

In severe acidosis, the patient may show cardiac arrhythmias, leading to possible cardiac arrest, due to hyperkalemia. Potassium moves from intracellular fluid to extracellular fluid if cellular damage results from the acid state. The consequent hyperkalemia is compounded by the inability to secrete normal amounts of potassium ions since the secretory mechanism is clogged by excess hydrogen ions.

Metabolic acidosis *facilitates/inhibits* (circle one) nerve function.

QUESTION 131

inhibits

ANSWER

198 REGULATION OF HYDROGEN ION CONCENTRATION

QUESTION 132 — The rapid and deep respirations observed in the patient with metabolic acidosis are attributable to the effort of the _____ system to _____.

ANSWER — respiratory ... compensate

QUESTION 133 — Cardiac arrhythmias accompanying metabolic acidosis are directly due to _____.

ANSWER — hyperkalemia

QUESTION 134 — The mutual aggravation of the conditions of metabolic acidosis and hyperkalemia is the outcome of the competition of _____ and _____ for renal secretory sites in the sodium exchange mechanism.

ANSWER — hydrogen ... potassium (in either order)

QUESTION 135 — Signs of the depressant effects of metabolic acidosis on nerve function include _____, _____, _____, and _____.

ANSWER — apathy ... disorientation ... stupor ... coma (any order)

ITEM 44 LABORATORY VALUES IN METABOLIC ACIDOSIS

As the kidneys attempt to compensate for metabolic acidosis, urinary pH drops below 6.0, unless the initial fault was the kidneys' inability to secrete hydrogen ions. Whereas serum bicarbonate levels initially may be below 25 mM./L., renal retention of bicarbonate tends to elevate HCO_3^- levels as compensation returns concentrations toward normal. The particular combination of values will vary with each patient, with the disease, and with the stage of compensation. Accurate information in any given instance can best be determined by serum electrolyte and arterial blood gas studies.

QUESTION 136 — In metabolic acidosis, serum bicarbonate levels are initially *high/low* (circle one).

ANSWER — low

Because of variation from case to case and in stage of compensation, reliable laboratory values for patients with metabolic acidosis should be determined by studies of serum _____ and arterial _____.

QUESTION 137

electrolytes ... blood gases

ANSWER

CAUSES OF METABOLIC ACIDOSIS

ITEM 45

Metabolic acidosis is a relative excess of hydrogen ions or deficit in sodium bicarbonate due to any cause other than respiratory pathology. Excess hydrogen ions accumulate from any condition which causes protein and fat, in preference to carbohydrates, to be metabolized for energy. These conditions include diabetes mellitus, prolonged fasting or vomiting, starvation, infections accompanied by fever, and hyperthyroidism. With drastic increases in hydrogen ion accumulation, the kidney is simply unable to secrete the excess.

In the list below, place a check mark in front of the items which are factors in metabolic acidosis.

( ) 1. increased H_2CO_3
( ) 2. decreased H_2CO_3
( ) 3. increased metabolic acids
( ) 4. decreased metabolic acids
( ) 5. increased HCO_3^-
( ) 6. decreased HCO_3^-

QUESTION 138

You should have marks in front of Nos. 3 and 6, increased metabolic acids and decreased bicarbonate.

ANSWER

The metabolic shift which results in H^+ accumulation and thereby causes metabolic acidosis is the utilization of _____ and _____ for energy.

QUESTION 139

protein ... fat

ANSWER

In addition to fasting, starving, and vomiting, three pathological states which may engender metabolic acidosis are _____, _____, and _____.

QUESTION 140

fever ... diabetes ... hyperthyroidism (any order)

ANSWER

ITEM 46 METABOLIC ACIDOSIS DUE TO BICARBONATE DEFICIT

Acidosis is caused not only by an increase in H^+ but also by a decrease in HCO_3^-. Since pancreatic secretions, intestinal secretions, and bile are all alkaline, prolonged diarrhea effects a total loss of base. Normally, this loss of base would be partially reabsorbed in the large intestine. Vomiting, too, can lead to loss of base—in spite of the fact that vomiting initially tends to remove gastric hydrochloric acid, leaving a base excess. What happens is that, eventually, vomiting lead to loss of intestinal contents, depleting the alkaline supply and leaving an acid excess.

QUESTION 141   Bicarbonate deficit leading to metabolic acidosis results when diarrhea depletes alkaline _____ secretions, _____ secretions, and _____.

ANSWER   pancreatic ... intestinal ... bile

QUESTION 142   Short periods of vomiting involve loss of gastric hydrochloric acid, creating a tendency to _____, whereas prolonged vomiting includes the loss also of alkaline intestinal secretions, generating a tendency to _____.

ANSWER   alkalosis ... acidosis

ITEM 47 LACTIC ACIDOSIS

Lactic acidosis results from any condition which causes severe tissue anoxia. Lactic acid is an intermediate metabolic product in the Krebs tricarboxylic acid cycle. Since the final stages of carbohydrate metabolism are aerobic (requiring oxygen), an oxygen deficiency at the tissue level will impede the completion of the metabolic process, resulting in lactic acid accumulation and subsequent acidosis.

QUESTION 143   Severe tissue _____ leads to accumulation of lactic acid, bringing on a condition of metabolic acidosis.

ANSWER   anoxia

Lactic acid is an intermediate metabolite in the _____ acid cycle.

QUESTION 144

Krebs tricarboxylic

ANSWER

CAUSES OF LACTIC ACIDOSIS

ITEM 48

Any condition which impairs oxygen transport is a potential cause of excess build-up of lactic acid. Conspicuous among such conditions are anemia, decompensated congestive heart failure, and hepatic disease. Prognosis is poorest, however, when the tissue anoxia results from some form of pulmonary dysfunction which impairs alveolar ventilation—as in emphysema, bronchiectasis, or any chronic obstructive pulmonary disease. In these cases, the inability of the lungs to compensate means that in effect, a respiratory acidosis is superimposed on a metabolic acidosis.

Disease conditions other than those of a pulmonary nature which may eventuate in lactic acidosis include _____, _____ disease, and uncompensated _____.

QUESTION 145

anemia ... hepatic ... congestive heart failure

ANSWER

Lactic acidosis due to the impaired oxygen transport in conditions of pulmonary dysfunction, such as emphysema and bronchiectasis, actually involves a combined _____ and _____ acidosis.

QUESTION 146

respiratory ... metabolic

ANSWER

COMPENSATION FOR METABOLIC ACIDOSIS

ITEM 49

When metabolic acidosis occurs, the lungs attempt to compensate by increasing alveolar ventilation in order to "blow off" excess CO_2 and lower H^+ levels. Respirations are deep and rapid. If the cause of the acidosis is not renal insufficiency itself, the kidneys aid by stepping up the rate of exchange of sodium for hydrogen and by increasing ammonia synthesis. Reabsorbed Na⁺ combines with retained HCO_3^-, thus raising the base component of the buffer system and restoring the acidotic body fluids to normal. Urinary pH is acid.

QUESTION 147
Pulmonary compensation for metabolic acidosis consists of increased _____ exchange.

ANSWER
respiratory

QUESTION 148
The kidneys remove excess acids by (1) secretion of _____ ions and (2) _____ synthesis.

ANSWER
hydrogen . . . ammonia

QUESTION 149
As hydrogen ions are secreted, the kidneys reabsorb _____, which combines with _____ to raise the base component of the buffer system.

ANSWER
sodium . . . bicarbonate

ITEM 50 TREATMENT OF METABOLIC ACIDOSIS

Treatment of metabolic acidosis obviously centers on correcting the causative lesion. Meanwhile, the patient's laboratory values must be restored to normal. If uncontrolled diabetes, vomiting, or diarrhea has resulted in dehydration, an intravenous fluid replacement of normal saline or Ringer's lactate solution may be ordered. Intravenous sodium bicarbonate or sodium lactate may be administered to correct a base bicarbonate deficit. If hyperkalemia is present, potassium intake should be curtailed and ion exchange resin enemas employed. As treatment proceeds cautiously, the patient is observed for symptoms of rebound metabolic alkalosis.

QUESTION 150
In treating metabolic acidosis, IV fluid replacement is aimed at correcting possible _____ as well as at correcting the _____ itself.

ANSWER
dehydration . . . acidosis

In treating metabolic acidosis, potassium intake may be limited and an ion exchange resin administered by enema to combat a concomitantly existing _____.	QUESTION 151
hyperkalemia	ANSWER

SIGNS AND SYMPTOMS OF METABOLIC ALKALOSIS   ITEM 51

Metabolic alkalosis affects primarily the neuromuscular system, the lungs, and the heart. The patient may appear nervous, irritable, and disoriented because of the hyperirritability of nerve tissue in alkalosis. Patients who are prone to seizures may convulse as a result of irritable foci developing in the brain. Hypocalcemia, leading to tetany, may develop as a consequence of the decreased ionization of calcium compounds which accompanies alkalinity.

Besides nerve tissue, other tissues which are affected by metabolic alkalosis include _____, _____, and _____ tissue.	QUESTION 152
muscle . . . lung . . . heart	ANSWER
The nervousness and disorientation symptomatic of metabolic alkalosis are due to the _____ of _____ tissue.	QUESTION 153
hyperirritability . . . nerve	ANSWER
Metabolic alkalosis has an adverse effect on serum calcium levels because calcium is less _____ in a milieu of alkalinity.	QUESTION 154
ionized	ANSWER

LABORATORY VALUES IN METABOLIC ALKALOSIS   ITEM 52

The patient with metabolic alkalosis will show total CO_2 content elevated above 27 mM./L. and will elaborate an alkaline urine. If arterial pH is greater than 7.45 and pCO_2 is normal (between 35 and 45 mm. Hg), an uncompensated metabolic alkalosis is suggested.

If arterial pH is between 7.35 and 7.45 and pCO_2 is elevated, compensation for the metabolic alkalosis, by the respiratory system's retention of carbon dioxide, is suggested. The ratio of base to acid in the buffer pair is now normal, and pH is stable.

QUESTION 155

Uncompensated metabolic alkalosis is to be inferred when arterial pH is greater than _____ and pCO_2 is _____.

ANSWER

7.45 ... normal

QUESTION 156

Indications that the respiratory system has compensated for the metabolic alkalosis are an arterial pH between _____ and _____ and an _____ pCO_2.

ANSWER

7.35 ... 7.45 ... elevated

ITEM 53   CAUSES OF METABOLIC ALKALOSIS

Metabolic alkalosis results from an abnormally high base bicarbonate level of body fluids or from an abnormally low hydrogen ion concentration. A common cause of bicarbonate excess is overzealous use of sodium bicarbonate as a home remedy for peptic ulcer. Patients who are vomiting gastric contents or those on gastric suction are prone to metabolic alkalosis from loss of hydrochloric acid. Loss of hydrogen ions increases the relative ratio of base to acid. Loss of chloride alters the ionized Na^+/Cl^- component of body fluids, leaving sodium free to form further sodium bicarbonate.

QUESTION 157

Place a check mark before those items listed below that are causes of metabolic alkalosis.

( ) 1. increased H_2CO_3
( ) 2. decreased H_2CO_3
( ) 3. increased metabolic acids
( ) 4. decreased metabolic acids
( ) 5. increased HCO_3^-
( ) 6. decreased HCO_3^-

ANSWER

You should have checked Nos. 4 and 5, decreased metabolic acids and increased bicarbonate.

QUESTION 158

Metabolic alkalosis frequently results from ingestion of excessive quantities of _____.

ANSWER

sodium bicarbonate

Metabolic alkalosis can result from a loss of acid, as occurs in _____ and _____.	QUESTION 159
vomiting ... gastric suction	ANSWER
When chloride is lost in gastric suctioning, a condition of increased serum levels of _____ may ensue.	QUESTION 160
sodium bicarbonate	ANSWER

COMPENSATION FOR METABOLIC ALKALOSIS

ITEM 54

In the buffer compensation for metabolic alkalosis, excess sodium bicarbonate reacts with the acid component of all the other buffer systems, raising the base ratio of these other systems and thereby distributing the alkali load. In the respiratory compensation, respirations become slow and shallow as alveolar ventilation decreases in an attempt to conserve CO_2 and H^+. With decreased respiratory exchange, oxygenation is decreased—obviously an undesirable side effect. If pO_2 falls to a level of cellular hypoxia, then a rebound lactic acidosis may occur. If hypoxia occurs, the patient may become cyanotic.

When buffer systems compensate for alkalosis, the excess bicarbonate reacts with the _____ component of all the buffer systems.	QUESTION 161
acid	ANSWER
Respirations in a patient with metabolic alkalosis tend to be _____ and _____.	QUESTION 162
slow ... shallow (either order)	ANSWER
Respiratory compensation for metabolic alkalosis is aimed at retention of _____ and _____.	QUESTION 163
carbon dioxide ... hydrogen ions (CO_2 ... H^+)	ANSWER

206 REGULATION OF HYDROGEN ION CONCENTRATION

QUESTION 164 An unwanted side effect of respiratory compensation may be _____, leading to _____ acidosis and _____.

ANSWER hypoxia ... lactic ... cyanosis

ITEM 55 — RENAL COMPENSATION FOR METABOLIC ALKALOSIS

The kidneys attempt to conserve hydrogen ions and excrete sodium bicarbonate. Initially high levels of sodium cause the adrenal glands to decrease secretion of aldosterone and thereby slow down the exchange mechanism in which sodium is reabsorbed in exchange for secretion of hydrogen or potassium ions. Because hydrogen ions are relatively scarce in alkalosis, potassium ions are preferentially secreted, leading to possible hypokalemia.

In addition to slowing the excretion of H^+ through the exchange mechanism, the kidneys also slow ammonia synthesis, providing yet another means of conserving, rather than excreting, hydrogen ions.

QUESTION 165 In order to slow the sodium/hydrogen exchange mechanism, the adrenal glands reduce secretion of _____.

ANSWER aldosterone

QUESTION 166 Alkalosis tends toward _____, in that potassium ions tend to be _____ into glomerular filtrate because of the scarcity of _____ ions.

ANSWER hypokalemia ... secreted ... hydrogen

QUESTION 167 In order to conserve H^+ in body fluids, the kidneys decrease synthesis of _____.

ANSWER ammonia

METABOLIC ALKALOSIS AND HYPOKALEMIA

ITEM 56

If serum potassium levels drop below normal, a patient with metabolic alkalosis may show all the signs of hypokalemia, including arrhythmias (leading to possible cardiac arrest), hypotension, muscular weakness and flaccidity, and paralytic ileus. Laboratory findings show total CO_2 content, principally bicarbonate, elevated above 27 mM./L. in arterial blood; arterial pH above 7.4; and urinary pH above 7.0.

The metabolic alkalotic patient's serum potassium levels may drop below normal, compounding the alkalosis with a simultaneous _____.

QUESTION 168

hypokalemia

ANSWER

The most dangerous effect of hypokalemia on the patient with metabolic alkalosis is _____ leading to possible _____.

QUESTION 169

arrhythmias ... cardiac arrest

ANSWER

A patient with metabolic alkalosis will have

(1) total CO_2 content elevated above _____ mM./L.

(2) arterial pH elevated above _____

(3) urinary pH above _____

QUESTION 170

27 ... 7.4 ... 7.0

ANSWER

CORRECTION OF HYPOKALEMIA IN THE ALKALOTIC PATIENT

ITEM 57

Before the patient's metabolic alkalosis can be handled successfully, it is necessary to cope with his hypokalemia by increasing his potassium intake. Additional intake is especially urgent in view of the fact that the short supply of K^+ is being rapidly depleted through excretion in order that the body may conserve the even shorter supply of H^+. Replacement of K^+ through increased intake is also needed to overcome extracellular K^+ depletion, which causes a shift of potassium from within cells to the extracellular compartment. To maintain electrical neutrality as intracellular K⁺ shifts out of cells, extracellular sodium and hydrogen move into the cells. When continued long enough, this exchange of extracellular H⁺ and Na⁺ for intracellular K⁺ can result in an intracellular acidosis being added to the already existing extracellular alkalosis.

QUESTION 171   Either oral or intravenous _____ intake is the most effective expedient for treating hypokalemia in the alkalotic patient.

ANSWER   potassium

QUESTION 172   When potassium is depleted in extracellular fluids, some potassium diffuses from _____ _____ into the _____ compartment to help offset the deficit.

ANSWER   intracellular fluids . . . extracellular

QUESTION 173   When intracellular fluids lose potassium ions, extracellular sodium and hydrogen shift intracellularly, creating a tendency to intracellular _____ that develops concomitantly with extracellular alkalosis.

ANSWER   acidosis

ITEM 58 TREATMENT OF METABOLIC ALKALOSIS

Treatment of metabolic alkalosis begins with alleviating the condition responsible for the imbalance. As a preventive measure, gastric tubes can be irrigated with isotonic saline rather than plain water. If alkalosis is due to infusion of alkaline drugs, the drug regimen obviously must be changed. If it is due to loss of gastric hydrochloric acid by vomiting or suctioning, then replacement of electrolytes, especially of chloride and potassium, is indicated. Ringer's lactate is frequently given because it contains both potassium and chloride, as well as being a corrective for alkalosis. Normal saline may be used to replace chloride, with potassium chloride added as needed.

QUESTION 174   Two electrolytes which are in need of replacement in metabolic alkalosis due to gastric suctioning are _____ and _____.

ANSWER   chloride . . . potassium (either order)

Potassium and chloride can be replaced by intravenous administration of _____ _____ or by saline with _____ _____ added.

QUESTION 175

Ringer's lactate . . . potassium chloride

ANSWER

SUMMARY OF UNIT FOUR

Biochemical reactions are extremely sensitive to the concentration of hydrogen ions in body fluids. The concentration of hydrogen ions per liter is represented by the symbol pH, which means the negative logarithm of the hydrogen ion concentration. A low pH means a high concentration; a high pH means a low concentration. Arterial blood is normally stabilized at pH 7.4, and venous blood at pH 7.35. When arterial pH drops below 7.4, the patient is in a state of acidemia; when arterial pH is above 7.4, he is in a state of alkalemia. A pH below 7.0 or above 7.8 is lethal.

Since enzymatically catalyzed metabolic reactions are especially vulnerable to even slight alteration in pH, it follows that there must be a means of stabilizing hydrogen ion concentration in spite of forces which tend to offset the balance. Three mechanisms regulate hydrogen ion concentration or restore acid-base balance to normal when a condition of imbalance occurs. These mechanisms are (1) the buffer systems, (2) the respiratory system, and (3) renal mechanisms.

Body fluids contain several combinations of weak acids and their conjugate bases serving as buffers. By definition, a buffer is a solution of a weak acid and its conjugate base combined in such a way that the solution resists drastic change in pH when a strong acid or base is added. Strong acid combines with the buffer base to form the buffer acid, weaker than the original acid. Conversely, strong base combines with the buffer acid to form the buffer base, weaker than the original base. By this system, strong acids added to body fluids are converted to weak acids and strong bases added to body fluids are converted to weak bases, with little variation in pH.

The three principal buffers in body fluids are bicarbonate/carbonic acid, disodium phosphate/monosodium phosphate and proteinate/acid protein. Their effectiveness depends on the concentration of each buffer pair and whether the ratio of base/acid is optimal.

The second means of stabilizing pH is by varying the rate of removal of carbon dioxide. Carbon dioxide as a metabolic end-product combines with water to form carbonic acid. Increasing the carbon dioxide content of body fluids lowers pH by forming more H_2CO_3 and more H^+. Conversely, decreasing the carbon dioxide content of body fluids forms less H_2CO_3 and less H^+ and raises pH.

Increased hydrogen ion concentration stimulates the respiratory center in the medulla to increase both respiratory rate and depth, providing a negative feedback control. As hydrogen ions accumulate, respiratory exchange is increased, excess carbon dioxide is eliminated, excess H_2CO_3 is removed, hydrogen ion concentration is lowered, and pH increases to cut down on the original stimulus.

When, on the other hand, any pathologic condition prevents the respiratory system from carrying out its normal negative feedback control, hydrogen ions build up, causing acidosis of respiratory origin.

The kidneys regulate acid-base balance by (1) tubular secretion of hydrogen ions, (2) reabsorption of sodium ions, (3) conservation of bicarbonate, and (4) synthesis of ammonia. By varying reabsorption and secretion of these substances across renal epithelium, the kidneys remove excess acid or base from body fluids by excreting it into urine.

Cells in the proximal tubules, distal tubules, and collecting ducts remove hydrogen ions by secreting them into the tubular lumen in exchange for sodium. Sodium combines with bicarbonate and diffuses into extracellular fluids. The net effect is removal of hydrogen ions from body fluids into urine and retention of sodium bicarbonate.

Hydrogen ion secretion continues against a concentration gradient and above the level at which bicarbonate is removed by reaction with urinary buffers and combination with ammonia. Buffers in urine are the bicarbonate and phosphate systems. Ammonia synthesized by tubular epithelium reacts with additional hydrogen ions forming ammonium, which then combines with chloride.

Renal mechanisms compensate for respiratory imbalance and respiratory mechanisms partially compensate for imbalance due to kidney malfunction. Respiratory acidosis is compensated for by an increased acidification of urine, effected by secretion of hydrogen ions, retention of bicarbonate, and synthesis of ammonia. Respiratory alkalosis is compensated for by a decrease in the above mechanisms, and excretion of additional quantities of bicarbonate. Metabolic acidosis is partially compensated for by hyperventilation and metabolic alkalosis by hypoventilation.

Acidosis and alkalosis occur as clinical entities when pathological states cause the body's acid-base regulatory mechanisms to be exceeded. Respiratory acidosis is due to retention of carbon dioxide in respiratory insufficiency. Respiratory alkalosis is caused by excess carbon dioxide "blow-off" in hyperventilation. Acidosis or alkalosis from any cause other than respiratory pathology is referred to as metabolic.

The signs of respiratory acidosis include dyspnea, cyanosis, and accelerated pulse rate. Pathological conditions commonly producing respiratory acidosis are pneumonia, emphysema, asthma, bronchiectasis, pulmonary edema, and depression of the respiratory center in the medulla. The body attempts to compensate partially by exchanging extracellular chloride for intracellular bicarbonate. The kidneys attempt to retain HCO_3^-, excrete H^+, and accelerate ammonia synthesis. Clinical intervention includes supportive therapy to alleviate the pulmonary disease causing the acidosis, IV infusion of sodium bicarbonate or Ringer's lactate, and gastric suction to remove hydrochloric acid from the stomach. Two complications sometimes encountered in connection with respiratory acidosis are hyperkalemia and hypocalcemia. The hyperkalemia is due to the preferential secretion of H^+ and retention of K^+ as hydrogen and potassium compete for the secretory sites in the sodium exchange mechanism. Hypocalcemia may occur because of the development of a rebound metabolic alkalosis in the patient with respiratory acidosis. Potassium excretion may be promoted by diuretics. Calcium insufficiency may be managed with calcium chloride.

Signs and symptoms of respiratory alkalosis are (1) neuromuscular irritability, hyperreflexia and twitching, (2) arterial pCO_2 below normal range, and (3) an alkaline urine. Respiratory alkalosis accompanies hyperventilation and certain effects of drugs, such as aspirin poisoning. To compensate, the kidneys attempt to excrete sodium bicarbonate, retain H^+, and reduce ammonia synthesis. A particular complication which may accompany respiratory alkalosis is hypokalemia, which can develop when excessive potassium ions are excreted in order to conserve hydrogen ions.

Symptoms observed in metabolic acidosis are apathy, disorientation, and other signs of depressed nerve function. Respirations are deep and rapid as the lungs attempt to compensate. In severe cases, arrhythmias appear which may lead to cardiac arrest. Urinary pH may drop below 6.0. Metabolic acidosis is commonly brought on by diabetes, hyperthyroidism, vomiting, starvation, and infections accompanied by fever. Prolonged diarrhea may deplete alkaline secretions, causing acidosis as a bicarbonate deficit develops. Tissue anoxia from any condition which impairs oxygen transport, such as anemia, may cause accumulation of lactic acid and thereby provide another source of metabolic acidosis. Both the lungs and kidneys attempt to compensate for metabolic acidosis, the former by hyperventilation, the latter by secreting H^+ in exchange for Na^+ reabsorption and by increasing ammonia synthesis to promote removal of H^+. Treatment of metabolic acidosis is aimed ultimately at removing the initial disease condition which has caused it. IV infusion of sodium bicarbonate is employed to correct bicarbonate deficit. If the patient has concomitant hyperkalemia, potassium intake must be restricted. Treating metabolic acidosis must proceed cautiously in order to forestall a rebound metabolic alkalosis.

The nervous symptoms of metabolic alkalosis are the opposite of those of metabolic acidosis. While the acidotic person shows signs of depressed nerve function, the alkalotic patient appears irritable—possibly convulsive—and manifests other signs of the effects of hyperirritability on the neuromuscular system, lungs, and heart. Because calcium compounds do not ionize readily in an alkaline environment, the metabolic alkalotic patient may be prone to hypocalcemia. Total CO_2 content in metabolic alkalosis is elevated above 27 mM./L., pCO_2 values may be normal, and arterial pH is greater than 7.45. To compensate for metabolic alkalosis, excess sodium bicarbonate reacts with the acid component of all other buffer systems to distribute the alkali load. Respirations slow down to conserve CO_2 and H^+. High sodium levels reduce aldosterone secretion, thus slowing the excretion of H^+ and reabsorption of Na^+ in the sodium exchange mechanism. The kidneys further promote conservation of H^+ by reducing ammonia synthesis.

A special complication may accompany metabolic alkalosis—namely, the development of hypokalemia because of the preferential secretion of potassium ions given the scarcity of hydrogen ions. Before the patient's metabolic acidosis can be treated successfully, it is first necessary to deal with the hypokalemia, by increasing potassium intake to overcome the K^+ deficit. Therapy for the metabolic alkalosis itself is directed at eliminating the original cause of the alkalotic condition.

QUESTION 176

Regulation of acid-base balance in body fluids is important because enzymatically catalyzed biochemical reactions are particularly vulnerable to the slightest change in _____.

ANSWER

pH

QUESTION 177

The three mechanisms which stabilize the concentration of hydrogen ions in body fluids are

(1) _____ systems

(2) _____ system

(3) _____ mechanisms

ANSWER

buffer ... respiratory ... renal

QUESTION 178

An increase in hydrogen ion concentration in a solution _____ pH; a decrease in hydrogen ion concentration _____ pH.

ANSWER

lowers ... raises

REGULATION OF HYDROGEN ION CONCENTRATION

QUESTION 179 A buffer, by definition, is a solution of a weak _____ and the salt of that _____ combined in such a way that the solution resists drastic change in _____ when a strong _____ or a strong _____ is added to the solution.

ANSWER acid ... acid ... pH ... acid ... base

QUESTION 180 Buffers are effective in converting _____ acids to _____ acids by reaction of the _____ _____ with the buffer _____.

ANSWER strong ... weak ... strong acid ... base

QUESTION 181 The respiratory system is effective in relieving acid stress by removal of _____ _____, which effectively takes acid out of extracellular fluids.

ANSWER carbon dioxide

QUESTION 182 Increased hydrogen ion concentration stimulates the _____ center in the medulla to effect an _____ in alveolar ventilation.

ANSWER respiratory ... increase

QUESTION 183 As the respiratory system responds to increased hydrogen ion concentration and as pH is _____, the increase in alveolar ventilation _____.

ANSWER stabilized ... diminishes

QUESTION 184 The kidneys regulate acid-base balance by

(1) tubular secretion of _____ _____

(2) reabsorption of _____ _____

(3) conservation of _____

(4) synthesis of _____

ANSWER
(1) hydrogen ions
(2) sodium ions
(3) bicarbonate
(4) ammonia

TREATMENT OF METABOLIC ALKALOSIS

QUESTION 185
Renal mechanisms compensate for _____ imbalances and _____ mechanisms compensate for renal malfunction.

ANSWER
respiratory ... respiratory

QUESTION 186
When the body's normal mechanisms for stabilizing pH are exceeded by acid or alkali stress, _____ or _____ results.

ANSWER
acidosis ... alkalosis

QUESTION 187
Respiratory acidosis is caused by retention of _____.

ANSWER
carbon dioxide

QUESTION 188
Respiratory alkalosis is caused by excessive removal of _____.

ANSWER
carbon dioxide

QUESTION 189
Metabolic acidosis results from a decrease in the ratio of base/acid in body fluids due to any cause other than _____.

ANSWER
respiratory pathology

QUESTION 190
Metabolic alkalosis results from an increase in the ratio of _____ to _____ in body fluids due to any cause other than _____.

ANSWER
base ... acid ... respiratory pathology

QUESTION 191
Clinically, respiratory acidosis accompanies _____, _____, and depression of the _____ center.

ANSWER
pneumonia ... emphysema ... respiratory
(asthma ... bronchiectasis ... pulmonary edema)

REGULATION OF HYDROGEN ION CONCENTRATION

QUESTION 192 Dyspnea, cyanosis, and accelerated pulse rate are symptomatic of respiratory *acidosis/alkalosis* (circle one); hyperreflexia, twitching, and irritability are symptomatic of respiratory *acidosis/alkalosis* (circle one).

ANSWER acidosis ... alkalosis

QUESTION 193 One device by which the body attempts to compensate for respiratory acidosis is the exchange of extracellular _____ for intracellular bicarbonate.

ANSWER chloride

QUESTION 194 The kidneys attempt to compensate for respiratory acidosis by _____ HCO_3^-, _____ H^+, and _____ ammonia synthesis.

ANSWER retaining ... secreting ... increasing

QUESTION 195 Hyperkalemia sometimes develops in the respiratory acidotic patient because in the competition for secretory sites in the sodium exchange mechanism, _____ ions are preferentially secreted and _____ ions retained.

ANSWER hydrogen ... potassium

QUESTION 196 Hypocalcemia may occur in the patient with respiratory acidosis because of a rebound _____ alkalosis.

ANSWER metabolic

QUESTION 197 Two forms of clinical intervention employed in respiratory acidosis are IV infusion of _____ _____ and gastric suction to remove _____ from the stomach.

ANSWER sodium bicarbonate (Ringer's lactate) ... hydrochloric acid

QUESTION 198 Respiratory alkalosis accompanies _____.

ANSWER hyperventilation

TREATMENT OF METABOLIC ALKALOSIS 215

In respiratory alkalosis, arterial pCO_2 is *above/below* (circle one) normal, and the kidneys excrete an *acid/alkaline* (circle one) urine.

QUESTION 199

below ... alkaline

ANSWER

To compensate for respiratory alkalosis, the kidneys increase secretion of _____ _____; decrease secretion of _____ _____; and decrease _____ _____.

QUESTION 200

sodium bicarbonate ... hydrogen ions ... ammonia synthesis

ANSWER

Hypokalemia sometimes accompanies respiratory alkalosis because the kidneys excrete an excessive quantity of _____ ions in an effort to conserve _____ ions.

QUESTION 201

potassium ... hydrogen

ANSWER

Metabolic acidosis accompanies _____, _____, and _____.

QUESTION 202

diabetes ... diarrhea ... vomiting

ANSWER

The apathy and disorientation symptomatic of metabolic acidosis result from depressed _____ function.

QUESTION 203

nerve

ANSWER

Tissue anoxia may contribute to metabolic acidosis by causing the accumulation of an intermediate product of metabolism, _____ acid.

QUESTION 204

lactic

ANSWER

Careful monitoring is necessary in treating metabolic acidosis if rebound _____ is to be prevented.

QUESTION 205

metabolic alkalosis

ANSWER

REGULATION OF HYDROGEN ION CONCENTRATION

QUESTION 206 Metabolic alkalosis produces *depression/hyperirritability* (circle one) of nerve function in the affected patient.

ANSWER hyperirritability

QUESTION 207 In metabolic alkalosis, total CO_2 content is *lower/higher* (circle one) than 27 mM./L. and arterial pH is *greater/less* (circle one) than 7.45.

ANSWER higher . . . greater

QUESTION 208 Hypocalcemia may sometimes accompany metabolic alkalosis because _____ compounds do not ionize readily in an _____ environment.

ANSWER calcium . . . alkaline

QUESTION 209 Hypokalemia may sometimes accompany metabolic alkalosis because the scarcity of _____ ions causes _____ ions to be preferentially secreted.

ANSWER hydrogen . . . potassium

QUESTION 210 Therapy for acidosis or alkalosis is primarily directed at removing the inciting pathology; until this is possible, acid-base balance can be restored by _____ therapy.

ANSWER replacement

UNIT FIVE • CLINICAL CORRELATIONS

DISEASE CONDITIONS DISRUPTING HOMEOSTASIS

ITEM 1

Almost any disease condition, if prolonged or severe, presents a problem in fluid and electrolyte balance because of its potential for disrupting one or several aspects of the homeostatic mechanism.

Disruption of homeostasis usually begins with some clinical entity unrelated in itself to fluid and electrolyte disturbance, such as an injury, malfunction of one of the organ systems, or bacterial invasion. Fluid and electrolyte problems arise secondary to this sort of original stress. Yet these metabolic consequences often pose an even greater threat to the patient's survival than the initial disease factor which brings them on. When any one organ system suffers loss of function it can no longer carry its share in the total regulation of the body's "internal environment." Additional stress placed on the remaining systems, if too great, compromises homeostasis dangerously.

This unit examines some of the more prevalent disease states which act as primary causes of metabolic upheaval. Specifically, this unit treats the following clinical conditions insofar as they relate to producing fluid and electrolyte imbalances: (1) postoperative states, (2) congestive heart failure, (3) liver disease, (4) renal insufficiency, (5) diabetes, (6) emphysema, and (7) burns. These seven conditions produce typical fluid and electrolyte problems. For example, heart failure typically is associated with edema, and so on. It must be remembered, however, that fluids and electrolytes and acid-base balance are so intimately interrelated that it is almost impossible to disturb one without implicating the others. It is therefore only in a rather relative and restricted sense that we can speak of each of the conditions treated in this unit as producing its own characteristic homeostatic disturbance. Each example will ordinarily initiate a whole chain of imbalances.

To sum up the relationship of disease conditions to homeostatic disruption, usually a pathological state totally unrelated to fluid and electrolytes occurs as a PRIMARY FACTOR (CAUSE). Problems of imbalance then arise as a SECONDARY FACTOR (RESULT). These secondary ill-effects may be a variety of interconnected forms of FLUID IMBALANCE, ELECTROLYTE IMBALANCE, and ACID-BASE IMBALANCE. The following diagram illustrates these relationships:

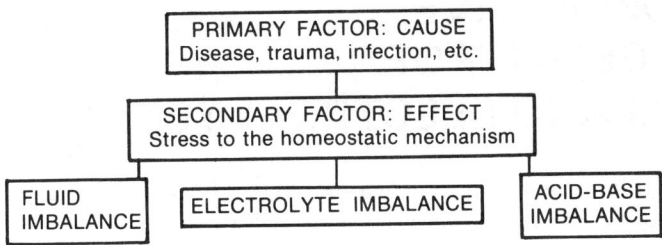

Many disease entities cause fluid and electrolyte problems by disrupting _____.

QUESTION 1

---

homeostasis

ANSWER

| QUESTION 2 | All organ systems share the responsibility for maintaining _____ and _____ balance. |

| ANSWER | fluid ... electrolyte |

| QUESTION 3 | When one system breaks down, it imposes an undue stress on other organ systems to preserve an optimum _____ environment. |

| ANSWER | internal |

| QUESTION 4 | Fluid and electrolyte imbalance usually develops secondarily to some primary _____ condition. |

| ANSWER | pathological |

| QUESTION 5 | Although a single disease condition may produce a single specific fluid and electrolyte disturbance, the usual situation involves problems involving several _____ organ systems. |

| ANSWER | interrelated |

ITEM 2 POSTOPERATIVE MANAGEMENT OF FLUIDS AND ELECTROLYTES

In the postoperative phase, management of the fluid, electrolyte, and acid-base status of the surgical patient focuses on many factors. (Prior to surgery, there is a period of planned hydration and optimum nutrition, followed by an NPO regimen—nothing by mouth—to lessen the chance of vomiting and aspiration.) The metabolic response after surgery follows a series of adjustments in fluid, electrolyte, and acid-base balance which represent part of the mechanism of compensation for tissue trauma. At times, however, a normal reflex response can itself create problems. For example, one of the post-traumatic adjustments to hemorrhage is retention of salt and water for the purpose of restoring blood volume and maintaining perfusion pressure. Necessary as this reflex response is for survival, it will require extremely specialized management if the patient's cardiac reserve happens to be already depleted. Although restorative reflex responses must be supported under normal conditions, enlightened therapeutics recognizes special situations in which these reflex responses require redirection.

The topics treated in the following items will be confined to problems of fluid, electrolyte, and acid-base balance in the immediate post-surgical period. Other aspects of post-surgical management, such as management of shock, for example, will not be taken up.

Post-surgical fluid and electrolyte adjustments are aimed at compensation for _____ _____. | QUESTION 6

---

tissue trauma | ANSWER

Normally restorative reflex responses to surgery may pose serious _____ to certain patients because of their particular physiological conditions. | QUESTION 7

---

problems (threats) | ANSWER

PREOPERATIVE INTRAVENOUS INFUSION | ITEM 3

Typically, a patient's homeostatic status is assessed prior to planned surgery, and any deficiency in electrolytes or nutrients is corrected in advance whenever possible. After an NPO regimen has been in force for several hours, the first intravenous solution is started upon entering the surgical area. Five per cent dextrose in water (D_5W), five per cent normal saline in water (D_5N/S), or Ringer's lactate (RL) may be ordered, depending on the patient's needs and the physician's preference. The purpose of the IV infusion is to maintain hydration, provide nutrition, and promote renal function. The presence of the IV line also provides a means of rapid administration of drugs, such as those which control blood pressure. Moreover, the IV line provides a route for blood transfusion should it be needed to support cardiovascular function and to avoid hypovolemia.

The primary purposes of the D_5W or other IV infusion prior to surgery are (1) maintenance of _____, (2) provision of _____ and (3) promotion of _____ function. | QUESTION 8

---

hydration ... nutrition ... renal | ANSWER

In addition to the primary purposes of the preoperative intravenous infusion, the IV line itself provides a readily available route for administration of needed _____ and for _____ _____. | QUESTION 9

---

drugs ... blood transfusions | ANSWER

ITEM 4 POST-SURGICAL SYMPTOMATOLOGY

During the first few days following surgical stress, there is a generalized protoplasmic catabolism that will shift to a protein anabolic phase as recovery proceeds. An increase in nitrogen excretion, creating a negative nitrogen balance, results partly from direct insult to cell and tissue integrity. A disruption of the restraining membranes enclosing intracellular protein releases nitrogenous substances into extracellular fluids for subsequent excretion. Quite often, the nitrogen excretion is disproportionately great in comparison to the degree of injury—that is, the negative nitrogen balance far exceeds that expected from membrane disruption alone. Its extent is magnified in proportion to the degree of endocrine and systemic alarm triggered by the tissue trauma.

Post-surgical symptomatology includes (1) weight loss due to the combined effects of protein and fat catabolism, a variable increase in basal metabolic rate, and low dietary intake; (2) retention of salt and water resulting from the release of aldosterone in response to stress, as well as antidiuretic hormone released in response to osmotic forces; (3) a period of oliguria, followed by a period of polyuria, which begins about the third day as earlier hormonal responses subside; and (4) various changes in electrolyte composition and distribution—specifically, sodium retention and total body potassium depletion due chiefly to an increase in aldosterone levels.

QUESTION 10 Negative nitrogen balance following surgery sometimes exceeds that expected from direct cell and tissue _____; the negative balance is magnified in proportion to the extent of the _____ reaction which accompanies tissue trauma.

ANSWER destruction ... stress

QUESTION 11 Specific electrolyte changes following surgery include retention of _____ and depletion of _____, due chiefly to an increase in _____ secretion.

ANSWER sodium ... potassium ... aldosterone

QUESTION 12 The following metabolic adjustments typically follow surgery:

(1) _____ loss

(2) retention of _____ and _____

(3) a period of _____ followed by a period of _____

(4) changes in electrolyte _____ and _____

ANSWER
(1) weight
(2) salt ... water
(3) oliguria ... polyuria
(4) composition ... distribution

When a patient loses weight following surgery, the causes are largely (1) a generalized protoplasmic _____, (2) an increase in _____ rate, and (3) a low _____ intake.	QUESTION 13
catabolism ... metabolic ... dietary	ANSWER

POSTOPERATIVE PROTEIN CATABOLISM ITEM 5

The body recuperating from the stress of surgery requires a source of additional energy. To supply this need, intensified glucocorticoid activity effects the formation of glucose from protein and fats, a process called *gluconeogenesis*. The protein catabolism involved in gluconeogenesis means that intact cells are hydrolyzing protein and utilizing it for energy. Increased excretion of nitrogen, most of it appearing as urea, is an index of protein catabolism following tissue destruction, as well as of protein mobilization from hormone fluctuations. When protein utilization and excretion exceeds protein synthesis in the body, the patient is in a state of negative nitrogen balance and possible metabolic acidosis. Infusion of a dextrose solution as an IV post-surgical replacement fluid helps provide needed glucose and reduces protein destruction.

Gluconeogenesis is a process of formation of glucose from _____ and _____.	QUESTION 14
fats ... protein	ANSWER
When protein is utilized for energy, _____ is released from the protein molecule and builds up in blood, mainly as _____.	QUESTION 15
nitrogen ... urea	ANSWER
When urinary nitrogen output exceeds nitrogen intake and synthesis, the patient is in _____ nitrogen balance, which means that body protein stores are being _____.	QUESTION 16
negative ... depleted	ANSWER

224 CLINICAL CORRELATIONS

QUESTION 17   The endocrine stress reaction is typically accompanied by _____ nitrogen balance because of _____.

ANSWER   negative ... gluconeogenesis

QUESTION 18   To decrease endogenous protein catabolism for energy, intravenous _____ is frequently given.

ANSWER   glucose (dextrose)

ITEM 6   ENDOCRINE RESPONSE FOLLOWING SURGERY

Fluid and electrolyte adjustments following surgery are brought on by the endocrine response to stress. Surgical trauma in some manner activates the pituitary gland to secrete adrenocorticotropic hormone (ACTH); ACTH, in turn, stimulates the adrenal cortex to release both hydrocortisone and aldosterone. Hydrocortisone modifies metabolic pathways so that fat and protein may be utilized for tissue repair as well as for a source of energy to supplement carbohydrates. Aldosterone increases salt and, indirectly, water retention, thus protecting vital structures against hypovolemia and decreased tissue perfusion.

Other endocrine responses occur secondarily. Antidiuretic hormone reestablishes the osmotic equilibrium disrupted by sodium retention. Increased norepinephrine and epinephrine secretion enhances heart rate and contractile force, affecting distribution of blood flow by constriction of certain vessels. Temperature increase with fever accelerates total metabolic rate and influences water and caloric balance, both of which are under endocrine control.

QUESTION 19   In response to the stress of tissue destruction, the _____ gland releases additional ACTH, and ACTH acts on the adrenal cortex to release _____ and _____.

ANSWER   pituitary ... hydrocortisone ... aldosterone

QUESTION 20   Utilization of fat and protein for tissue repair is mediated by *adrenocorticotropic hormone/hydrocortisone* (circle one).

ANSWER   hydrocortisone

In response to neurogenic stimuli, secretion of norepineprhine and epinephrine is increased in order to strengthen _____ and _____ force.	QUESTION 21

heart rate ... contractile ANSWER

POTASSIUM SHIFTS FOLLOWING SURGERY
ITEM 7

During the protein catabolic phase of postoperative recovery, potassium and water are released from the intracellular fluid of damaged tissue. Temporarily, serum potassium is increased and extracellular volume expanded. Since sodium is being retained because of high aldosterone levels, the excess serum potassium is promptly excreted, resulting in depletion of total body potassium. Even though the intracellular release of potassium is usually transitory and self-limiting, the potassium deficit needs replacement by judiciously regulated intake. Laboratory data on serum potassium levels will serve as an index of replacement needs required to maintain normal levels of 3.5 to 5.0 mEq./L. Forty to 120 mEq./24 hrs. of KCl may be routinely added to the immediate postoperative IV solution.

Once the phenomenon of cell release of potassium has passed, adequate oral intake ensures that (1) the intracellular supply will be replenished, (2) serum levels will stabilize, and (3) the functioning kidney will regulate total body potassium by excretion of excesses.

Since intracellular loss of potassium following surgical tissue trauma is accompanied by high levels of _____, the functioning kidney will excrete the excess _____. QUESTION 22

sodium ... potassium ANSWER

In the postoperative period, total body potassium is depleted through kidney excretion; accordingly, serum potassium levels should be monitored to determine intravenous _____ needs if the normal K^+ range of _____ to _____ mEq./L. is to be preserved. QUESTION 23

replacement ... 3.5 ... 5.0 ANSWER

ITEM 8 — FLUID SHIFTS FOLLOWING SURGERY

Release of water from intracellular fluids during postoperative recovery occurs secondarily to protein catabolism and potassium extrusion. During the protein catabolic phase, the shift of intracellular potassium to extracellular fluid reduces the osmotic pressure of intracellular fluid and increases the osmotic pressure of extracellular fluid. To maintain osmotic equilibrium, fluid shifts from the intracellular to the extracellular compartment. Extracellular solute and water concentrations increase at the expense of intracellular concentrations; extracellular mass expands as intracellular mass shrinks.

The temporary hypertonicity of extracellular fluids stimulates osmoreceptors to secrete more antidiuretic hormone. As ADH levels increase, additional water is reabsorbed from glomerular filtrate into blood. The role of urinary output as a mechanism for protecting extracellular fluid volume tends to decrease.

QUESTION 24 Protein catabolism and extrusion of intracellular potassium result in osmotic _____ shifts.

ANSWER fluid

QUESTION 25 During the protein catabolic phase of recovery, osmotic pressure increases in extracellular fluids because of increased _____ concentration.

ANSWER solute

QUESTION 26 Hypertonic extracellular fluid stimulates _____ to increase secretion of _____ hormone, causing reabsorption of water to protect _____ volume.

ANSWER osmoreceptors . . . antidiuretic . . . extracellular (blood)

ITEM 9 — STABILIZING BLOOD VOLUME FOLLOWING SURGERY

Extremely high blood volume is eventually attenuated by an increase in glomerular filtration rate and urinary output. In the initial series of postoperative events, however, endocrine factors primarily seem to dictate kidney response. Sodium and water are retained whether intake is high or low. Protective mechanisms aim at stabilizing blood volume and producing a slight edematous state, particularly at the operative site. Expansion of extracellular volume, then, occurs regardless of intake. Parenteral therapy is volume-regulated and geared to urinary output for the first few days in spite of already adequate blood volume. Otherwise, oliguria could progress to total anuria.

Increased blood volume is normally a stimulus to increase the rate of _____ and to increase _____ output.   QUESTION 27

glomerular filtration ... urinary   ANSWER

The volume of parenteral fluids is regulated to keep the patient in an *anuric/oliguric* (circle one) state during the first few days following surgery.   QUESTION 28

oliguric   ANSWER

In the post-surgical patient, factors which promote fluid *output/retention* (circle one) prevail over factors which promote fluid *output/retention* (circle one).   QUESTION 29

retention ... output   ANSWER

PROTEIN ANABOLISM AND POSITIVE NITROGEN BALANCE

ITEM 10

About a week or so following significant surgery, the adrenal reaction subsides and the anabolic phase of nitrogen metabolism begins. With an increase in dietary protein intake, urinary nitrogen excretion decreases. The patient's nitrogen balance shifts to positive as intracellular protein levels return to normal during the progress of convalescence. Along with the intracellular protein deposition, there is a return of intracellular potassium levels to normal. The explanation for the protein anabolic phase lies in the normalization of adrenal cortical hormone levels. The protein catabolic phase lasts as long as adrenal cortical activity remains high. With the subsiding of stress, glucocorticoid levels become lower, gluconeogenesis slows down, and carbohydrates again become the main energy source.

As the adrenal-pituitary stress following surgery begins to subside, protein *anabolism/ catabolism* (circle one) begins, and the patient's nitrogen balance shifts to *positive/ negative* (circle one).   QUESTION 30

anabolism ... positive   ANSWER

228 CLINICAL CORRELATIONS

QUESTION 31 Along with restoration of tissue protein, intracellular _____ returns to normal.

ANSWER potassium

QUESTION 32 Gluconeogenesis slows down as the body shifts from utilization of _____ and _____ to the utilization of _____ as a main energy source.

ANSWER proteins ... fats ... carbohydrates

ITEM 11 METABOLIC RESPONSE TO SURGERY—BRAIN AND HEART

Patient survival after surgery hinges on maintaining function in such vital organs as the brain, heart, lungs, and kidney. Several cardiovascular reflexes effect adequate tissue perfusion for protection of the brain and heart. Fluctuating blood volume is distributed in such a way that non-vital areas receive less blood in deference to the needs of the vital organs. Under the control of the autonomic nervous system, blood vessels constrict or dilate to regulate this distribution of blood flow.

As long as blood flows to vital areas, cellular homeostasis is maintained. That is, as long as the internal environment for individual neurons and individual myocardial fibers is adequate, the brain and heart can uphold the generalized homeostatic mechanism for all the other tissues of the body.

QUESTION 33 The vital organs whose functioning is essential to maintenance of homeostasis include the _____, _____, _____, and _____.

ANSWER brain ... heart ... lungs ... kidney (in any order)

QUESTION 34 Cardiovascular reflexes following surgery ensure flow of blood to vital areas in preference to non-vital areas by selective _____ and _____ of blood vessels.

ANSWER constriction ... dilation

METABOLIC RESPONSE TO SURGERY— KIDNEYS AND LUNGS

ITEM 12

The functioning of the kidneys and lungs following surgery is essential to the preservation of fluid, electrolyte, and acid-base balance. Electrolyte composition is determined almost entirely by the kidneys; the respiratory system shares with the kidneys the responsibility for acid-base balance. When either system is compromised, the other can temporarily compensate to some extent. When alveolar ventilation or kidney function is critically endangered, respirators and dialysis may be employed.

Electrolyte composition following surgery is controlled by the _____, while acid-base balance is maintained by both _____ and _____.

QUESTION 35

---

kidneys ... kidneys ... lungs

ANSWER

If both kidney and lung function should become compromised following surgery, mechanical compensation may be provided by _____ and _____.

QUESTION 36

---

dialysis ... respirators

ANSWER

POSTOPERATIVE ALVEOLAR VENTILATION AND ACIDOSIS

ITEM 13

When a patient has undergone thoracic or abdominal surgery, respiratory exchange may be decreased to the point of bringing on respiratory acidosis. If surgery involves the lung itself, or if a pulmonary embolus or atelectasis is superimposed on an already diminished respiratory reserve, the danger of respiratory acidosis is even more critical. The depressant action of anesthetics, as well as that of narcotics employed for post-surgical pain, can adversely affect respiratory exchange. A decrease in alveolar ventilation to the point of acidosis can also be provoked simply because of pain on breathing.

Decreased alveolar ventilation following surgery causes respiratory _____ from a build-up of _____ ions.

QUESTION 37

---

acidosis ... hydrogen

ANSWER

QUESTION 38

The use of drugs in connection with surgery can contribute to the development of respiratory acidosis because anesthetics and narcotics generally _____ respiratory function.

ANSWER

depress

QUESTION 39

An additional cause of carbon dioxide and hydrogen ion retention following thoracic and upper abdominal surgery is decreased _____ _____ due to _____ on breathing.

ANSWER

alveolar ventilation ... pain

ITEM 14 POSTOPERATIVE ALVEOLAR VENTILATION AND WATER BALANCE

The lungs play a significant role in water balance following surgery. Insensible water loss in the form of moisture in expired air accounts for 300 to 500 ml. of water loss in 24 hours. Decreased alveolar ventilation can slightly reduce this excretion of water, although not significantly. Of more importance is the increased water loss with fever. A febrile patient excretes more water in expired air because of his increased respiratory rate. The higher metabolic rate associated with fever also reduces total body water at a time when build-up of carbon dioxide is likely.

QUESTION 40

Alveolar ventilation affects water balance because of the moisture content in _____.

ANSWER

expired air

QUESTION 41

Insensible _____ _____ is diminished with decreased respiratory exchange but is increased with accelerated respiration, such as occurs with _____.

ANSWER

water loss ... fever

URINARY OUTPUT AND RENAL BLOOD FLOW FOLLOWING SURGERY

ITEM 15

Because of the rapid build-up of metabolic end-products immediately following surgery, continued kidney function is one of the most critical postoperative needs. Maintenance of glomerular filtration rate and urinary output is particularly critical for removal of nitrogenous substances, especially urea, which are released into blood by protein catabolism. Glomerular filtration rate depends on adequate pressure and flow in the renal artery; these factors in turn depend on cardiac output and blood volume. Sodium and water retention, promoted by increased secretion of aldosterone and antidiuretic hormone, protect blood volume. However, since the volume is fluctuating, the sympathetic nervous system controls constriction of afferent arterioles leading to the kidney, shunting blood away from the kidney to the brain and heart. Important as this neural reflex is to the brain and heart, it reduces renal blood flow and consequently decreases glomerular filtration rate and urinary output. The resulting renal ischemia, or lack of adequate blood flow for normal metabolic cell function, can cause extensive tubular necrosis, which, if unarrested, can lead to renal failure.

Immediately following surgery, there is generally an increase in blood urea _____.

QUESTION 42

---

nitrogen

ANSWER

Blood flow through the renal artery is frequently compromised by postoperative reflexes which divert blood away from the _____ to the _____ and _____.

QUESTION 43

---

kidney ... brain ... heart

ANSWER

Renal ischemia, which can bring on extensive _____ necrosis, results from inadequate renal _____.

QUESTION 44

---

tubular ... blood flow

ANSWER

POSTOPERATIVE TREATMENT OF REDUCED URINARY OUTPUT

ITEM 16

Oliguria or total anuria resulting from reduced renal blood flow—and aggravated by high circulating levels of ADH—must be detected as early as possible. When urinary output falls to about 300 ml. per day, inadequate renal blood flow should be suspected and prompt measures instituted to accelerate glomerular filtration rate. Intravenous fluids, set at a volume level to ensure glomerular filtration, are critical. Hydration may be geared so that the patient averages an hourly urine output of at least 30 ml./hour. If output drops below this level, either hydration is inadequate or renal failure is a possibility.

QUESTION 45

If urinary output falls below 300 ml. per day, a situation of inadequate renal _____ _____ is a serious possibility—one that calls for immediate clinical measures to speed up _____ rate.

ANSWER

blood flow . . . glomerular filtration

QUESTION 46

When the endocrine reaction to surgery tends to reduce glomerular filtration rate, the volume of intravenous fluids must be set at a level which will maintain _____ volume and renal _____ _____ .

ANSWER

blood . . . blood flow

ITEM 17 POSTOPERATIVE CASE HISTORY—DESCRIPTION OF THE CASE

Mrs. S., a 42-year-old housewife, was just beginning to regain consciousness as she was returned to her room after a cholecystectomy. A right subcostal incision had been made, the rectus abdominus muscle divided, and the peritoneum opened. The cystic duct and artery were ligated and sectioned, and the gallbladder was excised. A Penrose drain had been inserted and brought to the surface through a stab wound. Her stay in the recovery room had been uneventful except for an injection of morphine, gr. $1/6$, for pain. She arrived in her room with D_5 N/S infusing at 21 gtts/min. This initially hypertonic solution becomes isotonic as glucose is metabolized for energy. In addition to providing a direct source of energy, glucose allays the protein catabolic phase that follows surgical stress.

Surgery of Mrs. S.'s type temporarily reduces respiratory rate and depth, simply because the pain from the upper abdominal incision impedes normal breathing. Mrs. S. was assured that the IV infusion she was receiving was routine postoperative procedure in her case, and that she had come through the operation in fine shape. Her husband was allowed to visit as he felt necessary.

QUESTION 47

The intravenous solution administered to Mrs. S. following her surgery contained glucose, which served two purposes: (1) to provide a direct source of _____ and (2) to alleviate post-surgical _____ catabolism.

ANSWER

energy . . . protein

Mrs. S's diminished alveolar ventilation following surgery was attributable to the difficulty of breathing caused by _____.

QUESTION 48

pain

ANSWER

VITAL SIGNS AND POSTOPERATIVE CARE ON THE DAY OF THE OPERATION

ITEM 18

The first postoperative check of Mrs. S.'s vital signs showed the following values: blood pressure, 156/90; temperature, 98.4; pulse, 88; respirations, 24/min. Comparison of the values for these vital signs with preoperative baseline values suggested no particular need for concern, except in regard to improving the reduced, shallow pulmonary ventilation. For the remainder of the operating day, vital signs remained stabilized. Intake and output records were started.

The patient was positioned on her left side to prevent aspiration if she should vomit. She was required to turn, cough, and breathe deeply every two hours (with assistance). Morphine was given as needed for pain. Dressing checks showed no evidence of hemorrhage. There were signs, however, of bile leakage, so the dressing was reinforced.

The frequency with which the patient was required to turn, cough, and breathe deeply was every _____ _____.

QUESTION 49

two hours

ANSWER

POSTOPERATIVE CASE HISTORY— FIRST POSTOPERATIVE DAY

ITEM 19

On the morning of the first day following surgery, vital signs remained stable, differing only slightly from the day before. When Mrs. S. was reminded to cough and breathe deeply, she complained of pain. She continued to receive morphine, gr. $1/6$, as needed. The nurse showed her how to splint her side with a pillow to make coughing easier. She was assisted in deep breathing at those times when the analgesic's effect was at its maximum. She was instructed that deep breathing was critical, in that it would serve to prevent carbon dioxide build-up as well as to provide oxygen for tissue metabolism.

234 CLINICAL CORRELATIONS

Carbon dioxide and hydrogen ion retention carries the risk of respiratory acidosis. Hypoventilation also results in stasis of secretions.

When the doctor made his rounds, he told Mrs. S. she could be out of bed at least twice a day and had her dressing changed. That night, Mrs. S. slept soundly for about three hours and then awoke nauseated. Nausea continued off and on throughout the night. The nurse recorded a total output of vomitus 375 ml. Approximately 200 ml. of ice chips made of a balanced electrolyte solution were given to relieve thirst. Oral electrolyte solution in the form of ice chips is preferred to relieve thirst, because the addition of plain water with IV therapy creates the possibility of overhydration. Furthermore, plain water promotes movement of electrolytes into the stomach, making the solution isotonic, a condition to be avoided if there is any danger of vomiting or need for gastric suction.

At the end of the first postoperative day, intake was recorded at 3000 ml. D_5 N/S and approximately 500 ml. ice chips. Output was 1450 ml. urine, approximately 75 ml. drainage on the dressing, and 375 ml. vomitus. Her water balance, therefore, was positive.

QUESTION 50   Two disadvantages attend quenching thirst with plain water following IV therapy: (1) The patient may become _____ and (2) _____ may be lost.

ANSWER   overhydrated . . . electrolytes

QUESTION 51   If Mrs. S.'s vomiting were to continue over a longer period of time, it could lead to metabolic _____ as a result of loss of acidic gastric contents; or it could possibly lead to metabolic _____ because of loss of alkaline intestinal fluids.

ANSWER   alkalosis . . . acidosis

ITEM 20 POSTOPERATIVE CASE HISTORY— SECOND POSTOPERATIVE DAY

On the morning of the second day after surgery, Mrs. S. was served a surgical liquid breakfast since auscultation of the abdomen indicated that peristalsis had returned. Shortly after breakfast, however, she became nauseated. Vistaril was given for the nausea and she fell asleep. She awoke later in the morning with more nausea and vomiting. Vomiting continued intermittently throughout the second day and night. Vistaril was given every 3 to 4 hours.

Intake for the 24 hours was recorded at 3000 ml. D_5 N/S, 165 ml. apple juice, 75 ml. Jello, and about 300 ml. ice chips. Output was recorded at 1650 ml. urine, vomitus approximately 600 ml., drainage about 75 ml. The total intake of 3540 ml. when compared with the total output of 2325 ml. showed water balance still positive in spite of the vomiting.

The patient was still encouraged to cough and breathe deeply, but she found this very difficult to do on the second day. Blood was withdrawn for electrolyte determinations.

When water balance is positive following post-surgical stress, the patient is in *an oliguric/a diuretic* (circle one) phase.	QUESTION 52
an oliguric	ANSWER

POSTOPERATIVE CASE HISTORY— THIRD POSTOPERATIVE DAY — ITEM 21

On the morning of the third day, Mrs. S. refused her breakfast. She appeared weak, apathetic, and disoriented. Blood pressure was 110/72; temperature, 99.0; pulse, 90; respiration, 26/min. and shallow. Laboratory values showed hematocrit at 38; blood pH low at 7.34; total CO_2 content low at 22 mM./L.; Na^+ high at 152 mEq./L.; K^+, 4.1 mEq./L. Carbon dioxide pressure was high at 48 mm. Hg; pO_2 was 85 mm. Hg. Urinary pH was 5.0. BUN was normal at 18, indicating that the kidneys were functioning.

It was determined that Mrs. S. had a combined respiratory and metabolic acidosis. The respiratory acidosis was attributed to the shallow respirations consequent upon pain in breathing. The metabolic imbalance was due both to loss of bicarbonate with the vomiting of intestinal contents and to drainage of alkaline bile from the Penrose drain. Some degree of lactic acidosis was probably also present, because decreased alveolar ventilation would reduce the oxygenation at the tissue level.

Mrs. S.'s metabolic acidosis was brought on by loss of _____ in vomiting and loss of alkaline _____ through drainage.	QUESTION 53
bicarbonate . . . bile	ANSWER
Lactic acid accumulates and leads to acidosis when _____ is inadequate for tissue cell metabolism.	QUESTION 54
oxygenation	ANSWER

MANAGEMENT OF THE POSTOPERATIVE ACIDOSIS — ITEM 22

With a combined metabolic and respiratory acidosis, the only means of excreting excess hydrogen ions is through the kidneys. The conclusion was reached in Mrs. S.'s case that since renal function was intact, the problem would resolve itself with cessation of vomiting and drainage and with improved respiratory function.

236 CLINICAL CORRELATIONS

As a supportive measure, Mrs. S. was further encouraged to breathe deeply at those times when she was most comfortable. Her intravenous fluid was changed to lactated Ringer's solution. Whereas the glucose and saline solution has a pH range of 3.5 to 6.0, lactated Ringer's solution has a pH range of 6.0 to 7.5. In addition to providing sodium chloride, lactated Ringer's also supplies potassium chloride, calcium chloride, and sodium lactate. The lactate component supplies the alkalizing agent.

QUESTION 55
The doctor ordered Mrs. S.'s IV therapy changed to Ringer's lactate solution because of her low _____.

ANSWER
pH

ITEM 23
POSTOPERATIVE CASE STUDY— FINAL TREATMENT AND DISMISSAL

Late on the third postoperative day, as gastrointestinal function improved, Mrs. S. was able to resume the surgical liquid diet with no ill effects. The IV therapy was continued an additional day to maintain renal function and restore the total electrolyte balance. Total intake for the 24 hours of the third day was 3000 ml. Ringer's lactate and 1125 ml. oral intake. Urinary outut was 2770 ml., drainage slight, and vomitus negligible.

Although Mrs. S. still felt some pain when she took a deep breath, she was assured this would subside. The IV and the fluid balance studies were discontinued after the diuretic phase began. At this point, there was little danger of renal insufficiency occurring as a result of surgery. Mrs. S.'s continued recovery was uneventful. The Penrose drain was removed on the fifth day. She was dismissed on a low-fat diet at the end of 10 days.

QUESTION 56
As the diuretic phase began, water balance changed from _____ to _____.

ANSWER
positive . . . negative

QUESTION 57
The reasons for continuing IV therapy with Mrs. S. for one additional day as her recovery was progressing was to maintain _____ _____ and to restore total _____ _____.

ANSWER
renal function . . . electrolyte balance

CONGESTIVE HEART FAILURE AND FLUID BALANCE

ITEM 24

Among the chronic disabilities causing fluid imbalance, congestive heart failure is perhaps the most common. Congestive heart failure, often referred to simply as *cardiac failure*, is the inability of the heart to pump enough blood to meet normal circulatory demands. Cardiac failure is something quite distinct from what is commonly called *heart attack*. *Heart attack*, or *myocardial infarction*, which may produce *heart failure*, refers to an acute blockage of a coronary artery with consequent damage to part of the heart muscle. Heart muscle is damaged because of deprivation of blood supply to that localized area of the heart. Our primary concern here is with cardiac failure, not with heart attack.

Cardiac failure stems from a number of causes: hypertension; some arrhythmias; myocardial infarction; structural deformity of the valves; pressure around the heart; thyrotoxicosis; thiamine deficiency; arterio-venous fistulae; and others. Whatever the contributing events, cardiac failure implies a syndrome arising from failure of the heart muscle to respond adequately to circulatory demands.

Inability of the heart to pump enough blood to meet circulatory demands is termed

A. heart failure

B. heart attack

QUESTION 58

The correct answer is A. Heart failure, or cardiac failure, or, more accurately, congestive heart failure, is the heart's inability to pump sufficient blood. Heart attack is blockage of a coronary artery causing damage to the heart muscle.

ANSWER

Heart muscle damage due to inadequate blood supply, usually because of a clot in a coronary artery, is termed _____.

QUESTION 59

---

myocardial infarction

ANSWER

NORMAL HEART FUNCTION

ITEM 25

The following definitions help to clarify the phenomenon of heart failure: (1) *cardiac output* is the quantity of blood pumped by the left ventricle in one minute; (2) *stroke volume* is the volume of blood ejected with a single contraction; (3) *heart rate* is the number of beats, or contractions, per minute. Cardiac output = stroke volume × heart rate. Normally, the heart reflexivity adjusts stroke volume or heart rate—or both—to meet the diverse demands imposed by everyday activity.

CLINICAL CORRELATIONS

Cardiac output critically depends upon venous return, that is, the quantity of returning blood entering the right atrium per minute. Normally the heart will routinely adjust its rate and strength of contraction to accommodate variations in venous return, and propel forward all the venous blood entering the right atrium in any given time span. When, for any reason, the heart is unable to handle the venous return, the onset of congestive heart failure occurs.

QUESTION 60 The volume of blood pumped by the heart with each contraction is called the _____ _____; the number of beats of the heart per minute is called the _____ _____; the total quantity of blood pumped each minute is the _____ _____.

ANSWER stroke volume . . . heart rate . . . cardiac output

QUESTION 61 The circulatory system accommodates to varying demands by altering _____ _____, _____ _____, or both.

ANSWER stroke volume . . . heart rate

QUESTION 62 Normally, the heart adapts _____ _____ in such a way as to propel forward all the blood presented by venous return.

ANSWER stroke volume

QUESTION 63 The onset of congestive heart failure occurs when _____ _____ exceeds cardiac output.

ANSWER venous return

ITEM 26 THE HEART'S CONTRACTILE STRENGTH

The heart's ability to eject blood into the systemic circulation depends on its ability to contract. *Starling's Law of the Heart* states that, within physiological limits, increasing the stretch of heart muscle fibers increases their contractile force. Increasing venous return fills the ventricle with a greater quantity of blood and stretches the fibers. As might be expected, the heart responds with strengthened contractile force, up to a point.

Increased venous return increases the volume of blood entering the heart during diastole, the period of relaxation of heart muscle. When the heart fills with an increased volume of blood, heart muscle fibers are stretched and respond with an increased force of contraction during systole, which is the period of contraction of heart muscle. Increased contractile force, then, results in increased stroke volume.

Increasing the stretch of heart muscle fibers increases their _____ _____. QUESTION 64

contractile strength ANSWER

When the ventricles fill with extra quantities of blood owing to increased _____ _____, heart muscle responds, up to a point, by increasing its _____ _____. QUESTION 65

venous return . . . contractile strength ANSWER

IMPAIRMENT OF THE CONTRACTILE STRENGTH OF THE HEART

ITEM 27

If the heart rate holds stationary while the stroke volume increases, or vice versa, the heart will increase the total volume that it pumps per minute. This phenomenon is normal, up to a point, and constitutes a reserve to meet increased demands for blood. Eventually, however, the heart muscle reaches the limit of its ability to increase contractile strength. If excessive demands on cardiac output continue, contractile strength grows progressively weaker and cardiac output reaches a point of diminishing return. The early phase preceding clinical heart failure has been reached. From this point, the physiological limits set forth in *Starling's Law* have been exceeded.

What now happens is that the weakened heart muscle, already stretched beyond its maximum endurance, ceases to propel blood forward as effectively as it must. Increasing quantities pool in the venous circulation. Right atrial pressure rises as systolic ejection fails to respond to diastolic filling. Cardiac output can eventually decrease to the point that perfusion of tissues is inadequate.

The cardiac failure described in the preceding paragraphs soon turns into a problem in fluid and electrolyte balance, notably pulmonary edema, systemic edema, and sodium retention.

When heart muscle fibers are stretched beyond physiological limits, contractility decreases, stroke volume decreases, and _____ _____ exceeds _____ _____. QUESTION 66

venous return . . . cardiac output ANSWER

QUESTION 67 When cardiac output cannot keep up with circulatory demands, or when cardiac output cannot keep up with _____ _____, congestive heart failure has occurred.

ANSWER venous return

QUESTION 68 Fluid and electrolyte problems which typically accompany congestive heart failure include principally

(1) _____edema

(2) _____edema

(3) _____retention

ANSWER
(1) pulmonary
(2) systemic
(3) sodium

ITEM 28 CARDIAC FAILURE AND PULMONARY EDEMA

The general designation *congestive heart failure* can be specified more precisely as *left heart failure* or *right heart failure*, depending on the side of the heart where the trouble begins. The most immediate consequence of left heart failure, as far as homeostasis is concerned, is pulmonary edema, or accumulation of fluid in the interstitial space of the lungs.

The process of developing pulmonary edema starts when the left ventricle, the most frequent site of infarction damage, fails to propel enough blood into the arterial circulation. This creates a back pressure in the lungs as venous blood continues to flow toward the heart. The failing left ventricle increases the back pressure first in the left atrium, then in the pulmonary vessels, and finally in the right heart. The right ventricle responds to increased diastolic filling by contracting harder. This increases the hydrostatic pressure of blood entering the pulmonary vessels, causing fluid to leak from serum into the interstitial spaces surrounding the alveolar sacs and thereby producing pulmonary edema. Because of increased fluid accumulation in lung tissue, diffusion of oxygen and carbon dioxide across respiratory membranes is impaired, leading to dyspnea.

QUESTION 69 Left ventricular failure increases the back pressure in the left _____, pulmonary _____, and right side of the _____ as blood is dammed up in these areas.

ANSWER atrium ... vessels ... heart

Initially, additional stretch on the fibers of the right heart causes it to _____ the strength of contractions.	QUESTION 70
increase	ANSWER
Increased hydrostatic pressure in the pulmonary vessels causes fluid leakage from _____ into _____ _____, and pulmonary _____ follows.	QUESTION 71
plasma ... interstitial space ... edema	ANSWER

CARDIAC FAILURE AND SYSTEMIC EDEMA

ITEM 29

While left heart failure brings on pulmonary edema, right heart failure produces systemic edema, a generalized accumulation of fluid in interstitial spaces throughout the body. Actually, right heart failure frequently results when a left heart failure imposes such additional workload on the right heart that the right ventricle is weakened.

With failure of the right ventricle, back pressure builds up in the right atrium, in the venous circulation, and in the systemic capillaries. Distention of the capillary beds increases capillary hydrostatic pressure, which, in turn, promotes diffusion of fluid into interstitial spaces, producing systemic edema.

When cardiac output falls in cardiac failure, heart rate accelerates to maintain blood pressure. At the same time, the kidneys reflexively reduce glomerular filtration in an effort to increase blood pressure by increasing blood volume. Hypervolemia ensues and aggravates the edema formation, causing a generalized systemic fluid retention. The systemic edema is most obvious in dependent positions because of gravity.

Prolonged strain on the right heart imposed by left heart failure eventually causes _____ _____ failure and an increase in blood pressure in the _____ circulation.	QUESTION 72
right heart ... venous	ANSWER
Reduced cardiac output initiates reflexes which cause the kidneys to _____ _____ in order to maintain blood _____.	QUESTION 73
retain water ... volume	ANSWER

| QUESTION 74 | Increased venous pressure causes increased _____ hydrostatic pressure, which promotes _____ of _____ into interstitial space. |

| ANSWER | capillary ... diffusion ... fluid |

ITEM 30 CARDIAC FAILURE AND SODIUM RETENTION

Sodium retention, a protective mechanism following congestive heart failure, poses a serious problem of generalized edema formation. Low cardiac output stimulates the adrenal cortex to release aldosterone. Aldosterone promotes reabsorption of sodium, creating osmotic forces which bring about reabsorption of water. While the aldosterone-sodium reflex accomplishes its compensatory purpose of swelling arterial blood volume toward the hypervolemic state to increase cardiac output, the continued accumulation of water leads to systemic edema.

Although total body sodium rises, laboratory measurements may indicate, paradoxically, hyponatremia. The explanation of the hyponatremia is that serum sodium levels are diluted by the additional reabsorbed water.

| QUESTION 75 | Low cardiac output stimulates the adrenal _____ to secrete additional quantities of _____, which promotes reabsorption of _____ by the kidney. |

| ANSWER | cortex ... aldosterone ... sodium |

| QUESTION 76 | The aldosterone-release endocrine reflex (1) creates osmotic forces which promote reabsorption of _____, (2) increases blood _____, and (3) attempts to return cardiac _____ to normal. |

| ANSWER | water ... volume ... output |

| QUESTION 77 | Because of water reabsorption, the paradoxical situation often occurs in which laboratory data indicate _____ in spite of increased total body sodium; the apparent reduction of sodium levels is actually due to _____ by the reabsorbed water. |

| ANSWER | hyponatremia ... dilution |

TREATMENT OF CARDIAC FAILURE WITH DIGITALIS

ITEM 31

Treatment of congestive heart failure aims at restoring cardiac output and blood pressure to a level which will balance intake and output of fluids. Common measures that are employed include restriction of salt intake, use of digitalis therapy, and imposition of limits on the patient's activity.

Digitalis is universally prescribed for congestive failure. This drug increases the efficiency of the failing heart by strengthening its contractile force. Digitalis effects calcium influx into contractile proteins to make the fibers contract more effectively and to reduce the recovery time between contractions. Heart rate slows with circulatory improvement. Reflexes which promoted sodium and fluid reabsorption become less active. As increased mechanical efficiency raises cardiac output and glomerular filtration, diuresis of edema fluids follows. Improved circulatory dynamics also reduce venous pressure and alleviate both pulmonary and systemic edema.

Treatment of congestive heart failure aims at restoring cardiac output and blood pressure to a level at which _____ _____ balances _____.

QUESTION 78

---

fluid output ... fluid intake

ANSWER

The most universally accepted drug therapy for patients with heart failure is _____, which acts by increasing the _____ _____ of the heart.

QUESTION 79

---

digitalis ... contractile force

ANSWER

Reflexively, heart rate _____ as contractile force is improved; as sodium and fluid retention is reduced, edema is _____.

QUESTION 80

---

slows ... reduced

ANSWER

OTHER TREATMENT OF CARDIAC FAILURE

ITEM 32

Bed rest is encouraged for two reasons. First, the recumbent position obviously cuts down the heart's workload. Second, lying down favors diuresis by the flow of edema fluids from the lower extremities into the venous circulation. Although diuretics are routinely prescribed, they must be used judiciously because of their potential for causing

244 CLINICAL CORRELATIONS

electrolyte disturbances. Thiazide diuretics, such as Diuril, Esidrix, and Lasix, owe their diuretic effect to inhibition of reabsorption of sodium, potassium, chloride, bicarbonate, and water in the proximal tubules. Thiazides create a possibility of hypokalemia, which is especially dangerous to a patient in congestive failure. This danger occurs because hypokalemia reduces the dosage threshold for digitalis toxicity and exposes the patient to the serious threat of heart arrhythmias.

Restriction of salt intake is an important means of discouraging fluid retention. The average diet contains 5 to 15 grams salt. This must be reduced to a maximum of 5 grams or below. Fluid intake may also be restricted until urinary output returns to normal.

QUESTION 81 In addition to digitalization, the patient with heart failure usually requires restriction of dietary _____.

ANSWER sodium (or salt)

QUESTION 82 Bed rest is prescribed for cardiac failure patients for two reasons:
(1) _____
(2) _____

ANSWER
(1) to cut down the work load on the heart
(2) to facilitate diuresis

QUESTION 83 Great care has to be exercised in prescribing diuretics for heart patients because of the danger of _____.

ANSWER hypokalemia

ITEM 33
CONGESTIVE HEART FAILURE: CASE HISTORY— INITIAL TREATMENT

The case of Mr. B., age 58, is a typical example of a patient with a myocardial infarction resulting in congestive heart failure. The present episode began with crushing, severe substernal pain which radiated to the left arm and chin, with eventual syncope. Thirty minutes later, when the patient arrived at the hospital, he was pale, diaphoretic, and apprehensive. His arterial blood pressure was 102/80; temperature, 99.6; pulse rate, 105; and respiratory rate, 28/min. Oxygen therapy was initiated in the emergency room at 6 L./min. after which Mr. B. was taken to the intensive coronary care unit. The initial orders included absolute bed rest to reduce the workload of the heart and limitation of visitors to members of the immediate family. Morphine was ordered for pain and phenobarbital for sedation. Heparin was given as an anticoagulant. Oxygen was continued, and Mr. B. was placed on constant monitored EKG.

QUESTION 84

Initial treatment for Mr. B.'s congestive heart failure included:

(1) _____ therapy

(2) bed rest to _____ the _____ of the heart

(3) _____ for pain

(4) phenobarbital for _____

(5) heparin as an _____

ANSWER

(1) oxygen

(2) reduce . . . workload

(3) morphine

(4) sedation

(5) anticoagulant

CONGESTIVE HEART FAILURE: CASE HISTORY— AFTER 24 HOURS

ITEM 34

At the end of the first 24 hours, Mr. B. appeared to be out of immediate danger. Evaluation of the patient's status at this point included a review of the contributory role of residual damage from a moderately severe attack two years previously and assessment of damage to the heart by the present myocardial infarction. If a sufficiently large area of heart muscle is deprived of its blood supply, functional myocardium is replaced by non-functional connective tissue, with a consequent reduction of the contractile strength of the heart.

Blood samples were sent to the laboratory for serum enzyme studies to assess the extent of myocardial damage from the present attack. Laboratory findings showed elevation of the white count, sedimentation rate, SGOT (serum glutamic oxaloacetic transaminase), CPK (creatine phosphokinase), and LDH (lactic dehydrogenase). The degree of elevation of serum levels of these enzymes, released when myocardial cells are destroyed, indicates the extent of damage to heart muscle. Mr. B. was found to have sustained additional damage to the posterior wall of the left ventricle. The heart, accordingly, had a further reduced contractile strength due to the additional loss of functional myocardium.

Immediately after the present attack, cardiac function had been drastically reduced, with stroke volume and cardiac output both decreased. Diminished mean arterial pressure

246 CLINICAL CORRELATIONS

had stimulated the pressoreceptor reflex. Receptors which respond to a decrease in blood pressure are located in the aortic arch and carotid sinus. Activation of this mechanism had increased the number of impulses reaching the patient's cardioacceleratory center in the medulla. The center responded by increasing the heart rate. Activation of the patient's sympathetic nervous system also increased the force of contraction. Sympathetic constriction of peripheral vessels deflected blood to vital areas (the heart and the brain).

QUESTION 85   Mr. B.'s previous heart attack had left his heart with a diminished _____ _____ because of replacement of some functional myocardium with _____ _____.

ANSWER   contractile force ... connective tissue

QUESTION 86   The degree of damage to heart muscle following myocardial infarction is assessed on the basis of serum levels of several _____ which are released from _____ cells when these cells are destroyed.

ANSWER   enzymes ... myocardial

QUESTION 87   A decrease in mean arterial pressure activated the _____ reflex, which responded with generalized _____ nervous system discharge.

ANSWER   pressoreceptor ... sympathetic

QUESTION 88   Heart rate _____ with sympathetic stimulation; even the damaged heart muscle responded with increased _____ force in an effort to reestablish cardiac _____.

ANSWER   increased ... contractile ... output

ITEM 35 CONGESTIVE HEART FAILURE: CASE HISTORY— AFTER 10 DAYS

At the end of the tenth day, Mr. B.'s EKG recordings showed no evidence of additional occlusions. Complete inactivity was continued, so that the work load on the heart would be minimal until collateral circulation could invade the anoxic myocardium. The period of time from a week to ten days after heart failure is critical. The area of ischemic myocardium is approaching necrosis as a collateral blood supply is being established and fibrous connective tissue is beginning to invade the area. If the area is small enough and the repair is rapid enough, the prognosis is favorable. Several times Mr. B. complained of dyspnea and twice coughed up a pink frothy fluid, though cyanosis was no longer evident. Respiratory rales indicated that pulmonary edema fluid was forming.

Sodium was restricted to 250 mg/day. Fluid intake was reduced to 1800 ml. Urinary output was recorded as 525 ml. Laboratory values showed a decreased hematocrit and hyponatremia. In spite of the reduced sodium intake, fluid was being retained, probably because of increased aldosterone and maximal sodium reabsorption. Distended neck veins indicated retention of fluid and circulatory overload.

Although digitalis carries some risk of inducing arrhythmias in the patient with a recent heart attack, digitalization was indicated to forestall circulatory overload. (Many physicians use a preparation which is short-acting, in case an overdose should occur, and complete digitalization is delayed until dosage can be carefully ascertained.) Cedilanid was the drug of choice.

Frequent EKG interpretations were necessary to distinguish any indication of digitalis toxicity superimposed on the already prolonged QRS complex and depressed ST segment resulting from the occlusion. EKG monitoring at frequent intervals was continued throughout digitalization.

Lasix was prescribed for its natriuretic (sodium excreting) potency. This natriuretic effect increases urinary output by reducing renal reabsorption of sodium, with consequent reduction of water reabsorption. Daily serum potassium levels were monitored, so that the hypokalemia which frequently accompanies thiazide diuresis did not become severe. Hypokalemia is dangerous in patients who are taking digitalis because hypokalemia increases the likelihood of digitalis toxicity and can lead to cardiac arrhythmias.

Pulmonary edema was evidenced by the occasional periods of _____ and coughing up of a _____ fluid, in addition to detectable respiratory _____ .

QUESTION 89

---

dyspnea ... pink frothy ... rales

ANSWER

Accumulation of fluid in the pulmonary tissue spaces was due to increased _____ pressure of blood behind the damaged ventricle.

QUESTION 90

---

back

ANSWER

Digitalis was indicated, even though this drug carries with it the risk of _____ in the patient with recent myocardial infarction.

QUESTION 91

---

arrhythmias

ANSWER

248 CLINICAL CORRELATIONS

QUESTION 92 Digitalis reduces edema and promotes diuresis by increasing the _____ _____ of the heart.

ANSWER contractile strength

QUESTION 93 Increased blood pressure accompanying the increased mechanical efficiency of the digitalized heart increases glomerular _____ and promotes _____.

ANSWER filtration ... diuresis

QUESTION 94 Mr. B. was also given a thiazide diuretic to promote excretion of _____ and _____.

ANSWER sodium ... water

QUESTION 95 The natriuretic effect of thiazide diuretics can cause trouble with other ions, specifically, depletion of _____.

ANSWER potassium

QUESTION 96 Hypokalemia increases the possibility of digitalis _____ and its sequelae of cardiac _____.

ANSWER toxicity ... arrhythmias

ITEM 36 CONGESTIVE HEART FAILURE: CASE HISTORY— AFTER TWO WEEKS

At the end of two weeks, Mr. B.'s condition had improved enough to allow him to leave the intensive coronary care unit and continue his recovery in a private room. As active diuresis ensued, fluid balance was returned to normal with digitalization and a diuretic. Dosage was continued with digitoxin and Lasix. Potassium levels began to drop on the sixteenth day, but hypokalemia was avoided by giving KCl in 500 ml. of 5 per cent dextrose in water intravenously. Oral potassium (as K-Lyte) was prescribed and another diuretic was alternated periodically to reduce the tendency to hypokalemia.

Mr. B. continued an uneventful hospitalization and was released at the end of six weeks. Digitalis, once initiated, is unavoidably continued indefinitely. Dietary salt was restricted to 2 grams per day.

At the end of three months, Mr. B. returned to part-time work. EKG findings showed minimal damage, congestive heart failure was compensated for by reduced activity, and Mr. B. was asymptomatic.

As the congestive failure improved, _____ balance returned to normal.	QUESTION 97
fluid	ANSWER
The mechanical efficiency of the heart muscle increased as _____ was completed.	QUESTION 98
digitalization	ANSWER
Edema was relieved with increased efficiency of capillary _____ and increased urinary _____.	QUESTION 99
dynamics ... volume	ANSWER
Serum potassium levels were closely watched because of the tendency for _____ in a patient receiving diuretics.	QUESTION 100
hypokalemia	ANSWER
Hypokalemia is to be avoided in the digitalized patient because of its tendency to cause _____ toxicity and _____.	QUESTION 101
digitalis ... arrhythmias	ANSWER

LIVER DISEASE—ASCITES ITEM 37

In liver disease, functional liver cells, the parenchymal cells, are destroyed and replaced by connective tissue, creating a condition which impairs fluid and electrolyte balance. Factors resultant from liver disease which cause fluid imbalance are obstruction of blood flow and hypoproteinemia. A third factor, hormone changes which accompany liver disease, initiates electrolyte problems as well as disturbance of fluid balance.

250 CLINICAL CORRELATIONS

Obstruction of blood flow is easy to visualize if we recall the circulatory anatomy of the liver. The portal vein, which enters the inferior surface of the liver, serves as the channel through which venous drainage of the stomach and intestines is routed. Within liver tissue, the portal vein forms increasingly smaller branches, portal venules, which distribute themselves as sinusoids around hepatic cellular plates two cells thick. About 1000 ml. of blood flows through the portal vein each minute, distributing itself into the venous sinusoids surrounding the hepatic cellular plates.

When the liver is diseased, connective tissue infiltrates into the area and clogs the lumina of the sinusoids. Because of the tremendous blood flow perfusing the area, the increased resistance within the liver itself interrupts capillary dynamics in the vessels feeding in from the abdominal region. Specifically, this resistance increases back pressure in veins of the abdomen. Fluid leaks out of these abdominal veins and accumulates in the peritoneal cavity. Such accumulation of fluid in the abdominal cavity is called *ascites*.

QUESTION 102 The characteristic finding in liver disease is destruction of functional liver cells and replacement by _____ _____, creating a situation which results in _____ and _____ imbalance.

ANSWER connective tissue ... fluid ... electrolyte

QUESTION 103 One of the factors resultant from the infiltration of connective tissue is obstruction of _____ _____ through the liver itself.

ANSWER blood flow

QUESTION 104 A further consequence of connective tissue infiltration is an increase in the back pressure of veins in the abdomen, causing _____ of _____ into the peritoneal cavity and a condition called _____.

ANSWER leakage ... fluid ... ascites

ITEM 38 LIVER DISEASE—
HYPOPROTEINEMIA

In liver disease, impaired hepatic synthesis of serum proteins results in *hypoproteinemia*, a condition which contributes to fluid imbalance. In normal capillary dynamics, proteins provide part of the osmotic force exerting an inward pull of fluid into the venous end of the capillary. When low serum protein levels reduce this inward pull, generalized edema occurs.

All serum proteins except gamma globulin are synthesized by the liver. When the liver is diseased, it becomes unable to synthesize these substances in adequate quantities, with the result that production of albumin, fibrinogen, and alpha and beta globulin is seriously reduced.

The accumulation of ascites fluid accounts in large part for the hypoproteinemia that accompanies liver disease. The ascites fluid which accumulates in the abdominal cavity is extremely rich in albumin and electrolytes which have been absorbed from the intestine. Because the ascites fluid is hyperosmotic to body fluids, the protein in the ascites fluid remains lost from circulation, thus compounding the hypoproteinemia. The hyperosmotic pressure of ascites fluid also continues to draw additional water into the original volume, thus compounding the ascites condition.

One factor which contributes to fluid imbalance in liver disease is _____ resulting from impaired protein _____ and from loss of protein into _____ fluid.

QUESTION 105

---

hypoproteinemia . . . synthesis . . . ascites

ANSWER

Ascites, in itself, constitutes a displacement of fluid from the _____ system into the _____ cavity.

QUESTION 106

---

circulatory . . . abdominal

ANSWER

Ascites contributes to the generalized edema by displacement of _____, a situation which intensifies the effects of _____ on capillary dynamics.

QUESTION 107

---

protein . . . hypoproteinemia

ANSWER

LIVER DISEASE—
HORMONE ALTERATIONS

ITEM 39

For incompletely understood reasons, the circulating levels of antidiuretic hormone and aldosterone increase in liver disease. Whatever the reasons, an increase in these hormone levels contributes to fluid and electrolyte problems in the patient with liver disease.

Antidiuretic hormone increases the permeability of the distal convoluted tubules and collecting ducts of the kidney. An increase in antidiuretic hormone, then, increases the reabsorption of water from glomerular filtrate back into the extracellular fluids. As a result, urinary output is reduced, fluid is retained in the extracellular compartment of the body, and generalized edema occurs.

252 CLINICAL CORRELATIONS

Aldosterone increases reabsorption of sodium by the kidney, giving rise to a condition of hypernatremia, or increased sodium levels in the blood. Retained sodium directly initiates osmotic forces which cause water retention.

In addition to the water-retaining effect of hypernatremia, a potassium or hydrogen ion is secreted for every sodium ion reabsorbed across kidney tubular epithelium. Thus, an additional effect of sodium retention is the possibility of hypokalemia, which could lead to arrhythmias.

QUESTION 108 Hormonal causes of fluid retention in liver disease include increased _____ hormone and _____ secretion.

ANSWER antidiuretic . . . aldosterone

QUESTION 109 The effect of increased antidiuretic hormone is retention of _____, whereas the effect of increased aldosterone is retention of _____.

ANSWER water . . . sodium

QUESTION 110 As sodium is reabsorbed in larger quantities, not only is water retained by _____ forces, but, by exchange across the renal tubular epithelium, _____ is lost.

ANSWER osmotic . . . potassium

ITEM 40 LIVER DISEASE: CASE HISTORY—DIAGNOSIS

Mr. W. was brought into the emergency room by a policeman. He had been found wandering the streets in a state of mental confusion. The attending physician made the following observations: generally emaciated, abdomen distended, liver palpable and nodular, spleen enlarged and firm, muscles weak and flaccid, temperature 101.2°F., heart rate 120 per minute, breathing shallow. Edema was apparent in the lower extremities.

Mr. W. complained of abdominal discomfort but resisted the idea of hospitalization. Later questioning revealed the probability of alcoholism and chronic malnutrition, with dietary protein intake low, poor appetite, and occasional nausea and vomiting.

Laboratory tests for liver function confirmed the initial diagnosis. Bromsulphalein, a dye injection used as a test for liver function, was excessively retained in the circulatory system. This dye is destroyed by the liver and removed from serum rapidly when there is normal liver function. Mr. W. showed retention of 30 per cent of the original injection after 45 minutes, compared to normal retention of 5 per cent after 45 minutes.

Galactose tolerance test, another means of assaying the extent of functional liver cells, showed excretion of 18 grams of galactose (normal excretion is 3 grams) five hours after ingestion of 40 grams of galactose. This sugar is converted to glucose in a completely functional liver, so that very little galactose is excreted. Alkaline phosphatase and isocitric dehydrogenase levels were increased, indicating liver cell destruction and release of these enzymes into the blood.

Other laboratory tests showed serum sodium at 154 mEq./L. (normal is 140 mEq./L.), indicating hypernatremia due to sodium retention and dehydration. Serum potassium levels were low but in the normal range at 3.6 mEq./L. (the normal range is 3.5 to 5.0 mEq./L.). Serum albumin was 1.8 grams/100 ml. (normal 4.5 to 5.5 grams/100 ml.) and serum globulins were 3.0 grams/100 ml. (normal 1.5 to 3.0 grams/100 ml.). Diagnosis was alcohol-induced cirrhosis of the liver.

Emergency room examination showed the following symptoms indicative of liver disease:

(1) emaciation with muscles _____

(2) liver _____

(3) spleen _____

(4) _____ in the lower extremities

QUESTION 111

---

(1) flaccid

(2) palpable

(3) enlarged

(4) edema

ANSWER

Liver disease was confirmed by the Bromsulphalein test when a _____ quantity of this injected dye was retained in the circulatory system, indicating that the bromsulphalein had not been _____ by the liver, as would happen in normal liver function.

QUESTION 112

---

large . . . destroyed (or removed)

ANSWER

In the galactose tolerance test, a _____ quantity of this sugar is excreted by the diseased liver and a _____ quantity is converted to glucose.

QUESTION 113

---

large . . . small

ANSWER

Liver cell destruction is indicated when serum levels of the enzymes alkaline phosphate and isocitric dehydrogenase are *increased/decreased* (circle one).

QUESTION 114

---

increased

ANSWER

ITEM 41 LIVER DISEASE: CASE HISTORY—TREATMENT

Mr. W. was hospitalized and given complete bed rest. Dietary instructions included high protein intake (170 grams/day) and sodium restriction (less than 500 mg./day). Intravenous glucose and amino acids were initiated. Medication included a vitamin supplement rich in thiamine and methionine. Spironolactone, an aldosterone inhibitor, was given as a diuretic to combat the edema.

Recorded fluid intake was 2000 ml. at the end of the first 24 hours; recorded output was 2800 ml. as the edema fluid began to be mobilized.

At the end of 72 hours, visible traces of edema had disappeared, serum sodium had returned to normal, the ratio of serum albumin to globulin had improved, and fluid balance was stabilized. Palpable ascites fluid had diminished because of the improved circulatory dynamics associated with increased blood protein. The patient was rational and indicated a willingness to cooperate with the social worker assigned to his case.

Mr. W. continued to improve and was dismissed after five days with dietary instructions and caution concerning his alcohol intake. Medical intervention had slowed the course of events associated with cirrhosis and had restored fluid and electrolyte balance to normal.

QUESTION 115 In the hospital, Mr. W.'s dietary instructions included high _____ and _____ intake and low _____ intake.

ANSWER protein ... vitamin ... sodium

QUESTION 116 Spironolactone, an _____ inhibitor, was given as a _____.

ANSWER aldosterone ... diuretic

ITEM 42 GENERAL NATURE OF RENAL FAILURE

Renal failure is, literally, failure of the kidneys to perform their usual role in conservation and excretion of various substances. Specifically, the role of the kidney is excretion of water, excess electrolytes, and metabolic end-products. Excretion of these substances, with normal kidney function, is discriminative. Although metabolic wastes must be removed at a steady rate to prevent toxicity, the kidney must regulate the excretion of water and electrolytes very precisely to protect optimum extracellular levels from depletion. In other words, the kidney conserves only needed body water and electrolytes and eliminates the rest. Endocrine mechanisms regulate renal function to this end.

Kidney function is the sum total of the functional capacity of about two million nephrons. Although reserve capacity is great, there comes a point when progressive loss of functional nephrons reduces the regulatory abilities of the kidney. This is essentially what happens in renal insufficiency, or, as it is also called, renal failure.

Renal insufficiency results from loss of functional _____.

QUESTION 117

---

nephrons

ANSWER

Renal insufficiency means failure of the kidneys to perform their normal function of _____ and _____ of fluids and electrolytes and the removal of metabolic fluids and _____.

QUESTION 118

---

conservation ... excretion ... wastes

ANSWER

Normal kidney function removes metabolic wastes to prevent _____.

QUESTION 119

---

toxicity

ANSWER

Normal kidney function protects extracellular fluids from depletion of needed _____ and _____.

QUESTION 120

---

water ... electrolytes

ANSWER

Regulation of renal excretion of _____ and _____ is under control of _____ mechanisms.

QUESTION 121

---

water ... electrolytes ... endocrine

ANSWER

ACUTE AND CHRONIC RENAL FAILURE

ITEM 43

Chronic renal insufficiency presents a less favorable prognosis, although palliative treatment may prolong the patient's life for years. The prospects for recovery from chronic renal failure depend on whether the non-functional kidney tissue can return to a functional state. Chronic renal failure may have periods of acute flare-ups superimposed upon its underlying degenerative course. When the acute attack subsides, the course of the disease reverts to its former slow progression. In particularly insidious cases, so many nephrons may be destroyed that reserves are depleted before medical aid is sought.

Acute renal failure occurs secondarily to the decreased renal blood flow that accompanies shock or hypovolemia or as a concomitant of fluid loss in the patient with extensive burns. It also occurs in transfusion reactions as hemoglobin clogs the glomeruli, and it may be brought on by infections of the urinary tract. *Chronic renal failure* is a relentlessly progressive loss of functional nephrons due to renal ischemia, which can accompany congestive heart failure. Chronic renal failure may also occur with chronic pyelonephritis, glomerulonephritis, obstruction, and hypertension. Acute renal failure may subside spontaneously when the causative agent is removed, and treatment is aimed at keeping the composition of body fluids as nearly normal as possible until normal kidney function returns. Chronic renal failure implies permanent loss of function.

Many fluid and electrolyte problems are common to both acute and chronic renal failure: edema, acidosis, hyperkalemia, and azotemia (elevated non-protein nitrogen) leading to uremia (elevated blood urea levels).

QUESTION 122
Renal failure due to kidney damage from trauma, burns, transfusion reactions, and so on is called _____ renal failure.

ANSWER
acute

QUESTION 123
Progressive loss of functional nephrons due to pyelonephritis, glomerulonephritis, obstruction, hypertension, and so on, is termed _____ renal failure.

ANSWER
chronic

QUESTION 124
Treatment of acute renal failure aims at maintaining normal _____ and _____ balance.

ANSWER
fluid . . . electrolyte

QUESTION 125
Chronic renal failure has a more _____ course than acute renal failure, since more and more functioning _____ are destroyed.

ANSWER
progressive . . . nephrons

QUESTION 126
In addition to azotemia and uremia, fluid and electrolyte problems common to both acute and chronic renal failure include _____, _____, and _____.

ANSWER
edema . . . acidosis . . . hyperkalemia

RENAL FAILURE—REDUCED SERUM CLEARANCE AND SODIUM EXCRETION

ITEM 44

As the number of functional nephrons progressively decreases in chronic renal insufficiency, glomerular filtration rate decreases, renal blood flow is reduced, and the kidney shrinks in size. Reduced renal blood flow reduces the kidney's capacity for serum clearance, which means that the ability of the kidney to rid the body of toxic metabolites from serum decreases. Urea and other nitrogenous products build up to dangerous levels in the blood.

A special problem in renal insufficiency is the aggravation of edema when excess sodium builds up. When the insufficiency is due to reduced renal blood flow, glomerular filtration decreases. As the movement of filtrate is slowed in its course through the tubules, there is a tendency for excess sodium to be reabsorbed simply because of the length of time the filtrate remains in the tubules. When serum sodium levels rise too high, a diuretic is often needed to promote sodium excretion.

When renal insufficiency is due to renal ischemia and tubular necrosis, some of the functional parenchyma of the tubular cells is destroyed. In this case, all reabsorptive and secretive processes are impaired to whatever extent functional nephrons have been lost. In this situation, rather than reabsorbing too much sodium, the diseased kidney is unable either to conserve sodium or to secrete potassium and hydrogen ions.

In summary, patients with a decreased glomerular filtration rate have an impaired capacity to eliminate water and salt excess. On the other hand, patients with increased permeability of the glomerular membranes and loss of functional tubular cells will eliminate dangerously large quantities of water and salt. In every instance, sodium restriction as an aid to combating edema must be determined on an individual basis.

QUESTION 127

When chronic renal failure reduces the number of functional nephrons, both _____ _____ rate and renal _____ _____ are decreased.

ANSWER

glomerular filtration . . . blood flow

QUESTION 128

As plasma clearance is reduced in chronic renal failure, there is a build-up of _____ in the blood to _____ levels.

ANSWER

urea . . . toxic

QUESTION 129

With chronic renal failure, excretion of sodium may be

A. reduced

B. increased

C. either reduced or increased

ANSWER

The correct answer is C. If the chronic renal failure is due to decreased renal blood flow, sodium excretion is apt to be reduced; if the failure is due to loss of functional nephrons, sodium excretion is apt to be increased.

258 CLINICAL CORRELATIONS

| QUESTION 130 | Individualized diagnosis determines the advisability of _____ restriction for patients with renal failure. |

ANSWER: salt (or sodium)

ITEM 45 RENAL FAILURE AND HYPERKALEMIA

Hyperkalemia (high serum potassium levels) usually presents a greater threat in acute renal failure than in the chronic form of the disease. This is especially critical when tissue trauma has released large quantities of potassium from intracellular fluids. Potassium levels can, however, become dangerously high also in chronic cases and, in particular, during a superimposed crisis such as systemic illness with increased protein catabolism, injury, surgery, myocardial infarction, or any condition chracterized by cell destruction.

In both acute and chronic cases, acidosis accompanying the hyperkalemia markedly worsens the patient's condition, since both potassium and hydrogen ions compete for excretion in the sodium exchange mechanism. The seriousness of hyperkalemia-induced heart arrhythmias warrants careful attention to serum potassium levels any time kidney excretion mechanisms become impaired.

| QUESTION 131 | Hyperkalemia is usually more of a problem in _____ renal failure than in _____ renal failure. |

ANSWER: acute ... chronic

| QUESTION 132 | In acute renal failure, potassium levels are frequently elevated because of _____ _____ which has released large quantities of potassium from _____ fluids. |

ANSWER: tissue trauma ... intracellular

| QUESTION 133 | Hyperkalemia is aggravated when _____ complicates the patient's condition. |

ANSWER: acidosis

| QUESTION 134 | In all cases of impairment of kidney excretion mechanisms, serum _____ levels must be carefully monitored because of the danger of cardiac arrhythmias which may be induced by _____. |

ANSWER: potassium ... hyperkalemia

SYMPTOMS OF HYPERKALEMIA

ITEM 46

When plasma potassium levels rise above 5.0 mEq./L., the patient may begin to show signs and symptoms of hyperkalemia. A slight hyperkalemia may produce a neuromuscular irritability, for example, which manifests itself in intestinal colic and diarrhea. As the condition worsens, muscle weakness leading to flaccid paralysis may appear. Decreased neuromuscular sensitivity may bring on paralytic ileus. As the severity of the hyperkalemia increases, heart action is depressed, with resultant flaccidity, dilatation, and bradycardia.

A particualr complication of hyperkalemia is the tendency for the patient to develop acidosis as hydrogen ions are retained in increasing numbers because of the competition of K^+ and H^+ for secretory sites. Acidosis leads to depression of the central nervous system, with consequent disorientation, delirium, and coma. Respiratory rate and depth increase as the respiratory system attempts to compensate for the metabolic acidosis.

Among the more common signs of hyperkalemia are

(1) _____ disturbances

(2) _____ weakness

(3) _____ paralysis and paralytic _____

(4) decreased neuromuscular _____

(5) cardiac _____ and _____

(6) _____

QUESTION 135

---

(1) gastrointestinal
(2) muscle
(3) flaccid ... ileus
(4) irritability
(5) flaccidity ... dilatation
(6) bradycardia

ANSWER

The kidney patient who develops metabolic acidosis in addition to hyperkalemia will suffer from depression of the _____.

QUESTION 136

---

central nervous system

ANSWER

LABORATORY DETERMINATION OF SERUM POTASSIUM

ITEM 47

Paradoxically, many of the symptoms of hyperkalemia are the same as those of hypokalemia. For this reason, determination of serum potassium by laboratory means is mandatory.

As plasma potassium levels rise, typical electrocardiographic findings include high peaked T-waves, prolonged QRS-intervals, decreased amplitude of P-waves and R-waves, and depressed ST-segments. The situation is critical if serum potassium is greater than 8.0 mEq./L. or if an electrocardiograph reveals absence of P-waves or a broad QRS complex. Arrhythmias and cardiac arrest are a distinct possibility.

QUESTION 137   Electrocardiographic findings which suggest hyperkalemia include (1) _____ peaked T-waves, (2) _____ QRS-intervals, (3) _____ amplitude of P-waves and R-waves, and (4) _____ ST-segments.

ANSWER   high ... prolonged ... decreased ... depressed

QUESTION 138   Excess potassium levels in body fluids is critical when serum potassium exceeds _____ mEq./L., or when EKG findings include absent _____ or a broad _____.

ANSWER   8.0 ... P-waves ... QRS complex

QUESTION 139   Because symptoms of _____ and _____ can be the same, laboratory determination of serum potassium is mandatory.

ANSWER   hyperkalemia ... hypokalemia

ITEM 48 CLINICAL MANAGEMENT OF HYPERKALEMIA

It is common practice to combat hyperkalemia by the administration of calcium gluconate, sodium bicarbonate, or insulin. Calcium antagonizes the cardiac toxicity of hyperkalemia, especially if hypocalcemia is present, as it usually is in renal shutdown. Sodium bicarbonate combats acidosis and facilitates potassium secretion. Insulin promotes potassium reentry into cells, thus lowering the levels in blood.

When more drastic procedures are required to remove potassium from the body, peritoneal dialysis or hemodialysis is employed.

QUESTION 140   In lowering serum potassium levels, (1) calcium gluconate acts by antagonism of toxic _____ effects; (2) sodium bicarbonate acts by combating _____; (3) insulin acts by promoting _____ of potassium into _____.

ANSWER   cardiac ... acidosis ... reentry ... cells

When serum potassium levels are drastically high, the clinical device to be used is _____ _____ or _____.

QUESTION 141

---

peritoneal dialysis ... hemodialysis

ANSWER

RENAL FAILURE AND HYPOKALEMIA

ITEM 49

In contrast to the hyperkalemic symptomatology described in preceding items, some kidney patients show signs of hypokalemia, or low serum potassium levels. Hypokalemia occurs in renal failure when glomerular filtration is reduced and the progression of filtrate along the length of the tubules is slowed. Sodium is excessively reabsorbed, and potassium and hydrogen ions are excessively secreted. Vomiting and diarrhea typical of uremia frequently bring on hypokalemia. Potassium levels may also fall as a result of the administration of thiazide diuretics prescribed for edema. These diuretics have a sodium depleting activity, but they happen also to increase excretion of potassium as well as sodium.

Treatment of hypokalemia is more easily accomplished than is treatment of hyperkalemia. Many patients respond readily to increased potassium intake, provided in fruit juice, until kidney function adjusts excretion. Other patients may require potassium added to intravenous fluids.

Hypokalemia may result from the _____ and _____ which frequently accompany uremia.

QUESTION 142

---

vomiting ... diarrhea

ANSWER

Potassium is also depleted by administration of thiazide diuretics prescribed to combat _____.

QUESTION 143

---

edema

ANSWER

The specific factors which are conducive to hypokalemia in renal insufficiency are the slowing down of _____ _____ and the excessive reabsorption of _____ and secretion of _____.

QUESTION 144

---

glomerular filtration ... sodium ... potassium

ANSWER

QUESTION 145 Simple and usually effective remedies for hypokalemia in the kidney patient include dietary intake of _____ _____ and intravenous infusion of _____ supplement.

ANSWER fruit juices ... potassium

ITEM 50 — LABORATORY DETERMINATION OF HYPOKALEMIA

Recognizable signs and symptoms of hypokalemia include malaise, hyporeflexia, skeletal muscle atony, paresthesia, weak pulse, faint heart sounds, heart block, falling blood pressure, shallow respirations, and vomiting. Many symptoms of hypokalemia may be the same as those of hyperkalemia. All in all, potassium levels are dependent on too many variable factors to predict whether the patient with renal failure will be hypokalemic or hyperkalemic. In every case, it is necessary to rely finally on accurate and often repeated laboratory analyses of serum potassium levels.

Electrocardiographic signs of hypokalemia include a prolonged Q-T interval, depressed ST-segment, negative T-waves, A-V block, premature contractions, tachycardias, and atrial flutter and fibrillation. These findings occur when serum potassium falls from its normal range of about 3.5 to 5.0 mEq./L. to below 3.0 mEq./L.

QUESTION 146 In hypokalemia, an electrocardiogram will show the Q-T interval to be _____; the ST-segment _____; and T-waves _____.

ANSWER prolonged ... depressed ... negative

QUESTION 147 Electrocardiographic signs of hypokalemia appear when serum potassium levels fall below _____ mEq./L.

ANSWER 3.0

ITEM 51 — RENAL FAILURE—ACIDOSIS AND URINARY pH

A patient with renal insufficiency is usually acidotic, with urinary pH ranging near 5.0 as hydrogen ion concentration builds up. When urinary pH approaches 5.0, arterial pH nears a danger level. Remember that arterial pH should remain at 7.4. Anything below pH 7.0 or above 7.8 is incompatible with life.

Urinary pH varies more than arterial pH, either up or down. The reason for this is that the buffer capacity of blood exceeds the buffer capacity of urine. As hydrogen ions continue to build up in body fluids the patient becomes increasingly lethargic, with the danger of going into acidotic coma.

Impaired renal mechanisms for ridding the body of acid are not usually completely lost. As long as glomerular filtrate forms, it retains its usual buffer capacity. But demands for excretion become greater than the diseased kidney can accommodate with the continued progression of reduced glomerular filtration and renal blood flow.

Urinary pH can be misleading. Since pH measures free hydrogen ion concentration, it is difficult to say that a urinary pH of 5.0 shows a well-buffered acid urine, or whether the low pH simply indicates a partically buffered acid urine in which relatively few free hydrogen ions are causing a drastic pH reduction.

When acid-base balance in the patient with renal insufficiency is impaired, urinary pH fluctuates more than _____ pH.

QUESTION 148

arterial

ANSWER

The difference in arterial pH and urinary pH is due to the greater _____ capacity of the blood.

QUESTION 149

buffer

ANSWER

As the pH of body fluids drops, the patient becomes lethargic, and acidotic _____ becomes a real danger.

QUESTION 150

coma

ANSWER

As glomerular filtration becomes reduced in volume, the _____ capacity of glomerular _____ is impaired.

QUESTION 151

buffer ... filtrate

ANSWER

RENAL FAILURE AND AMMONIA SYNTHESIS

ITEM 52

The kidney has several mechanisms that effectively secrete acid from extracellular fluids into glomerular filtrate, which extends into excretion as an acid urine. One of these mechanisms is in the ability of the healthy kidney to synthesize ammonia. Ammonia combines with free hydrogen ions to form ammonium ions, thus "trapping" acid. Ammonium ions combine with chloride or sulfate in glomerular filtrate and are excreted as a neutral salt.

Renal failure grossly reduces ammonia synthesis, as a rule. This reduction is due to the fact that numerically fewer nephrons are operating. A prolonged acid load stimulates the normal renal epithelial cells to increase ammonia synthesis, rising to a peak after two or three days. In the diseased kidney, increase in ammonia synthesis in response to acid stimulus is considerably retarded.

QUESTION 152   Because a patient with renal failure is less able to synthesize _____ than the normal individual, that patient's capacity for hydrogen ion removal is _____.

ANSWER   ammonia . . . diminished

QUESTION 153   Ammonia traps acid by combining with free _____ ions to form _____ ions.

ANSWER   hydrogen . . . ammonium

QUESTION 154   Ammonium ions, when they combine with _____ or with _____ in glomerular filtrate, are excreted as neutral salt.

ANSWER   chloride . . . sulfate

ITEM 53   RENAL FAILURE AND HYDROGEN ION SECRETION

Hydrogen ions diffuse freely in peritubular fluids, following the direction imposed by concentration gradients. A concentration gradient favors diffusion into filtrate as long as filtrate buffers soak up free hydrogen ions. Since there is less buffering capacity in glomerular filtrate when glomerular filtration rate is low, the hydrogen ion "soaking-up" ability is low and a favorable concentration gradient no longer exists. At that point, diffusion of hydrogen ions into glomerular filtrate stops.

Build-up of free hydrogen ions inside the lumen effectively shuts down the concentration gradient which allowed diffusion from peritubular fluid to urine. Actually, the diffusion gradient operates at a new setting. The diseased kidney imposes a different level of acid-base control. The patient is usually slightly acidotic, with an arterial pH around 7.3 or 7.2 because of the lowered level of acid-base control. In such a situation the patient will not become severely acidotic until late in the course of the disease.

Hydrogen ions _____ freely in peritubular fluids.

QUESTION 155

diffuse

ANSWER

When the volume of glomerular filtrate decreases, its _____ ability is diminished, with the result that the _____ gradient necessary for diffusion to take place is decreased.

QUESTION 156

buffering ... concentration

ANSWER

With progressive loss of the kidney's ability to rid the body of excess hydrogen ions, pH of body fluids becomes progressively _____, until the patient becomes seriously _____ late in the course of the disease.

QUESTION 157

lower ... acidotic

ANSWER

NORMAL EXCRETION OF PROTEIN METABOLITES

ITEM 54

The end-products of protein metabolism, urea and creatinine, are both toxic at high blood concentrations. Serum concentration and glomerular filtration rate determine their rate of excretion. Urea and creatinine are neither actively reabsorbed nor actively secreted. They diffuse down a concentration gradient across glomerular and tubular membranes, with concentration on one side of the membrane equilibrating with concentrations on the other side. As glomerular filtrate is removed from the body in the form of urine, it takes with it a certain quantity of urea and creatinine.

As blood levels of urea and creatinine rise, the concentration gradient favors removal into glomerular filtrate. Decreased blood concentration of these metabolites, on the other hand, decreases the concentration gradient and slows removal. Thus an automatic system comes into play. Ordinarily, when there is increased protein intake, automatically increased diffusion will take care of the excess protein metabolites.

Urea and creatinine, end-products of _____ metabolism, are both toxic at high blood concentrations and must be continuously removed by the _____.

QUESTION 158

protein ... kidney

ANSWER

CLINICAL CORRELATIONS

QUESTION 159

The mechanism for removal of urea and creatinine is _____ down a _____ _____, a mechanism which is effective so long as glomerular filtrate continues to be formed.

ANSWER

diffusion ... concentration gradient

ITEM 55 — RENAL FAILURE AND EXCRETION OF PROTEIN METABOLITES

The patient with renal failure who is oliguric (having decreased urinary output) is removing fewer nitrogenous metabolites. As long as these substances continue to form from the ingestion of protein, their circulatory levels will rise. A serious problem appears when blood levels of urea rise at a rate faster than impaired kidney function can remove such metabolites. Toxic levels of urea, called *uremia*, lead eventually to coma if they go unchecked.

Creatinine is excreted at a relatively constant daily rate because it is primarily a product of endogenous tissue catabolism. Administration of carbohydrate and fat to spare protein from catabolism for energy can reduce serum creatinine. The demands on the kidney for excretion of urea, on the other hand, are increased with high protein intake. Simply decreasing dietary protein levels can delay dangerous levels of urea in patients with renal insufficiency. A similar situation can exist with water. When more water is ingested than the non-functional kidney can excrete, it is retained in body fluids and leads to edema. Both food intake and water intake may have to be regulated quite carefully in patients who are suffering from renal insufficiency.

QUESTION 160

An oliguric patient removes fewer _____ metabolites.

ANSWER

nitrogenous

QUESTION 161

If blood levels of urea and creatinine continue to rise while the non-functional kidney is producing no glomerular filtrate, the diffusion gradient is _____.

ANSWER

blocked

QUESTION 162

Build-up of urea to a toxic level in the blood is referred to as _____ and can lead to _____.

ANSWER

uremia ... coma

RENAL FAILURE 267

Toxic levels of urea can be forestalled in the oliguric patient by restriction of _____ intake; by the same token, edema can be forestalled by restricting the intake of _____ until output of urine resumes.

QUESTION 163

protein . . . water

ANSWER

EXACERBATION OF OTHER DISORDERS BY RENAL FAILURE

ITEM 56

Kidney failure often compounds problems arising from other disease entities. The kidney patient is more susceptible to infection. He becomes anemic as the diseased kidney produces less of the hormone *erythropoietin*, which stimulates bone marrow production of red blood cells. His bleeding tendency is increased because of a defect in platelet thromboplastic activity. Renal insufficiency complicates congestive heart failure with fluid retention and hypervolemia on the one hand, and, on the other, contributes to the cause of congestive failure by imposing an additional workload on a normal heart. Again, renal disease can both aggravate existing hypertension with hypervolemia as well as contribute to bringing hypertension on in the first place. Renal failure brings on hypertension by causing an abnormal release of renin. Renin, an enzyme produced by the kidney in response to low renal blood flow, activates angiotensin, a vasopressor substance in blood which produces hypertension.

The kidney patient is more susceptible to _____, is _____, and has a tendency to excessive _____.

QUESTION 164

infection . . . anemic . . . bleeding

ANSWER

Renal failure may bring on hypertension when low renal blood flow stimulates the kidney to produce the enzyme _____.

QUESTION 165

renin

ANSWER

TREATMENT OF RENAL FAILURE BY HEMODIALYSIS

ITEM 57

Theoretically, a patient with renal disease could be maintained indefinitely with dialysis. Both hemodialysis and peritoneal dialysis preserve optimum concentrations of extracellular fluids as a temporary measure, but both have limitations. Attaching a patient to an artificial kidney requires cannulation of two blood vessels. Since a vessel will eventually clot off, the measure is self-limiting as far as repeated use is concerned. Permanently implanted cannulae pose the difficulty of infection, although internal cannulae appear to be less troublesome.

In hemodialysis, blood is pumped from one vessel through the artificial kidney and then back into another vessel. While the blood is outside the body, it runs through tubing which is bathed in a dialysis fluid. This fluid contains glucose, water, and electrolytes in approximately the same concentrations as in body fluids, but it contains no urea nor creatinine. Urea and other wastes go from blood through the freely permeable tubing and into the dialysis fluid. Since the dialysis fluid has approximately the same concentration of water and electrolytes as body fluids, there is no net loss of needed constituents from the body as these substances move through the tubing.

If there is excessive build-up of certain electrolytes in body fluids, the composition of the dialysis fluid can be designed to remove them. If potassium, for example, is building to a dangerous level, this ion is decreased or omitted altogether from the dialysis fluid, allowing diffusion of potassium ions out of blood.

Glucose is added to the dialysis fluid to make the fluid hypertonic to body fluids so that no net transfer of water goes into the circulatory system. Indeed, glucose concentration may vary upward as a means of removing excess extracellular volume by osmosis to combat pulmonary and systemic edema.

QUESTION 166

The artificial kidney works on the principle of

A. diffusion

B. osmosis

ANSWER

The correct answer is A. Urea and other metabolic wastes, as well as excessive electrolytes, diffuse across the semipermeable tubing into the dialysis fluid. Indeed, one reason for adding glucose to the dialysis fluid is to prevent the fluid from moving by osmosis into the circulatory system.

QUESTION 167

Dialysis as a means of maintaining a patient with kidney disease poses problems over time because of _____ and _____ .

ANSWER

infection . . . clotting

QUESTION 168

In the process of hemodialysis, metabolites which must be removed from the body diffuse into the _____ fluid, but there is no net _____ of substances needed by the body.

ANSWER

dialysis . . . loss

RENAL FAILURE

TREATMENT OF RENAL FAILURE BY PERITONEAL DIALYSIS

ITEM 58

Peritoneal dialysis requires less sophisticated equipment than hemodialysis, but it is less effective and requires more time and more frequent repetition to lower levels below a point of toxicity. This procedure poses a danger of peritonitis, and, accordingly, utmost care must be taken that the technique is conducted under sterile conditions and with sterile solutions.

Peritoneal dialysis works on the same principle as hemodialysis, but the membranes of the peritoneum serve as the semipermeable membrane. An incision is made in the abdominal cavity. Fluid similar to extracellular fluids, but lacking the substances which must diffuse into it, is introduced into the cavity. After a period of time, equilibration of body fluids and dialysis fluid takes place. The composition of the dialysis fluid determines the rate and extent of diffusion of substances into it. After equilibration, the dialysis fluid is removed, taking with it the undesirable products which would otherwise rise to dangerous levels in the body.

Peritoneal dialysis, like the artificial kidney, works on the principle of _____, but in peritoneal dialysis the _____ serves as the semipermeable membrane.

QUESTION 169

---

diffusion . . . peritoneum

ANSWER

The main disadvantages of peritoneal dialysis are the need for _____ and the danger of _____.

QUESTION 170

---

repetition . . . peritonitis

ANSWER

In peritoneal dialysis, the dialysis fluid is allowed to flow into the _____ _____.

QUESTION 171

---

abdominal cavity

ANSWER

After a period of equilibration, the peritoneal dialysis fluid is withdrawn from the abdominal cavity, taking with it metabolic _____.

QUESTION 172

---

wastes

ANSWER

ITEM 59

RENAL FAILURE: CASE HISTORY— DESCRIPTION OF THE CASE

Mr. K., age 52, had occasional periods of nausea and vomiting recurring for several months. He also noticed increasing restlessness, fatigue, insomnia, and occasional pruritis. He had been treated by his family physician on three occasions within the last few years for a urinary infection. One day, alarmed at experiencing frequent and urgent urination, flank pain, and fever, he agreed to go into the hospital for tests. Mr. K.'s blood pressure was slightly elevated, tissue in the lower extremities edematous, and neck veins distended. The patient was slightly dyspneic. The odor of ammonia was apparent on his breath. Although the foregoing signs strongly suggested kidney involvement, further tests were indicated.

QUESTION 173

A part of Mr. K.'s medical history that might have suggested kidney problems was the fact that his family doctor had treated him several times for an infection of the _____ _____.

ANSWER

urinary tract

QUESTION 174

Mr. K.'s symptoms which suggested the possibility of kidney involvement included

(1) the odor of _____ on the breath

(2) distended _____ _____

(3) _____ in the lower extremities

(4) _____

ANSWER

(1) ammonia
(2) neck veins
(3) edema
(4) dyspnea

ITEM 60

RENAL FAILURE: CASE HISTORY— LABORATORY WORK-UP

Initial diagnostic work-up showed blood urea nitrogen (BUN) elevated at 62 mg./100 ml.; blood creatinine slightly elevated at 3 mg./100 ml.; serum sodium heightened at 150 mEq./L.; serum potassium elevated at 6.5 mEq./L.; serum calcium depressed at 4.4 mEq./L.; hematocrit decreased at 31 per cent; serum pH at 7.2; and urinary pH at 5.0.

The elevated BUN and creatinine levels indicated an impairment in glomerular filtration. Serum sodium and potassium were elevated because of the slower passage of filtrate through the tubules, permitting increased reabsorption in the proximal tubules. Secretion of potassium and hydrogen ions in the distal tubules was probably decreased, contributing to hyperkalemia and acidosis. Since serum calcium and phosphorus levels are inversely related, calcium was decreased because of impaired elimination of acid phosphates. Mr. K. was anemic as a result of impaired synthesis of erythropoietin.

A urinary bacterial count revealed more than 150 white blood cells per high power field (normal is 0 to 4); multiple casts; occasional red blood cells (normal is 0 to 3); and albuminuria. The specific gravity of the urine was high at 1.032, suggesting that fluid was being retained.

Mr. K. was diagnosed as having pyelonephritis, with a history of previous attacks. He was put on broad-spectrum antibiotic therapy until culture and sensitivity determinations could be made.

Elevated BUN and creatinine pointed to reduced _____ _____; elevated serum and potassium were accounted for by slower passage of _____ through the _____.

QUESTION 175

---

glomerular filtration ... filtrate ... tubules

ANSWER

Mr. K.'s depressed serum calcium levels were accounted for by reduced elimination of acid _____; his anemia resulted from failure to synthesize sufficient _____.

QUESTION 176

---

phosphates ... erythropoietin

ANSWER

RENAL FAILURE: CASE HISTORY— FIRST 24 HOURS

ITEM 61

Mr. K. was put on a high-fat, high-carbohydrate, low-protein diet to minimize exogenous and endogenous protein catabolism and to prevent a further increase in BUN. Sodium intake was restricted to 1.5 grams of salt per day to combat extracellular fluid retention and edema. Lasix was used to promote sodium and potassium excretion. During the first 24 hours, the patient alternated between periods of drowsiness and periods of irritability. He complained of difficulty in breathing and chest discomfort. Daily weighing and intake and output tabulations were begun to assess the course of the edema formation. In addition to receiving fluids orally, he was given one liter of an intravenous solution to combat acidosis. It was felt that an adequate fluid intake was needed to

272   CLINICAL CORRELATIONS

promote glomerular filtration and dilution of the infectious agent. Adequate fluid intake was also desired, as long as sufficient urinary output continued, as a means of lowering BUN. At the end of the first 24 hours, fluid intake was 2650 ml.; urinary output was 1375 ml. Breathing became less difficult as edema fluids began mobilizing and acidosis improved.

QUESTION 177   Mr. K.'s low-protein, high-fat diet was aimed at minimizing _____ _____ and arresting the increase in _____.

ANSWER   protein catabolism ... BUN

QUESTION 178   In spite of fluid retention, Mr. K.'s fluid intake was increased for three reasons:

(1) to promote _____ _____

(2) to _____ the infectious agent

(3) to lower _____

ANSWER   glomerular filtration ... dilute ... BUN

ITEM 62 RENAL FAILURE: CASE HISTORY— RESULTS OF LAMINAGRAPHY

The next morning, additional tests were begun. An x-ray of the kidneys, ureters, and bladder (KUB) revealed some cloudiness of the right kidney but no opacity indicative of calculi. A laminagram was called for in order to check the size of the kidney as a clue to whether the present infection was superimposed on a chronic renal insufficiency or whether it was due to an acute attack. Mr. K.'s right kidney was shrunken, indicating renal insufficiency of long standing. An intravenous pyelogram (IVP) confirmed a filling defect in the right kidney. Mr. K.'s doctor directed an extensive follow-up to protect the remaining functional nephrons.

QUESTION 179   A laminagram was used to determine the _____ of the kidney as an aid in diagnosing whether the kidney involvement was _____ or _____.

ANSWER   size ... acute ... chronic

The shrunken size of Mr. K.'s right kidney revealed that his renal insufficiency was *acute/chronic* (circle one).

QUESTION 180

---

chronic

ANSWER

To confirm the findings of the laminagram, Mr. K. was given an intravenous _____, which showed a _____ _____ in the right kidney.

QUESTION 181

---

pyelogram . . . filling defect

ANSWER

RENAL FAILURE: CASE HISTORY—DISMISSAL

ITEM 63

Bacterial count on the fourth day showed the infection subsiding, with edema fluid mobilizing and fluid balance negative. The IV was discontinued and fluid intake as desired was allowed the following day as kidney function resumed.

Lab values showed BUN at 25 mg./100 ml., serum sodium at 134 mEq./L., and serum potassium at 5.0 mEq./L. Continuing electrolyte checks over the next few days showed stable normal values.

The immediate danger past, Mr. K. was dismissed at the end of the week. The dotor instructed him to resume normal activity within a few days, and to eat a well-balanced, low-protein diet, with a salt substitute replacing table salt. Antibiotics were continued to combat any remaining infection, and diuretics were continued at a maintenance dose level. Mr. K. was instructed to return for a check-up at the end of two weeks. At this time, the fluid and electrolyte balance of the patient would be evaluated and any needed corrections in therapy made.

Increased urinary output indicated that _____ fluid was being mobilized with the diuretic.

QUESTION 182

---

edema

ANSWER

On dismissing Mr. K., the doctor outlined a well-balanced, _____ diet, with a _____ _____ to be used as a replacement for table salt.

QUESTION 183

---

low-protein . . . salt substitute

ANSWER

274 CLINICAL CORRELATIONS

ITEM 64 DIABETES—METABOLIC IMPLICATIONS

Diabetes mellitus is an endocrine disorder in which β-cells of the pancreas are unable to produce enough insulin for normal metabolism of carbohydrates. Insulin promotes transport of glucose across cell membranes for the enzymatic synthesis of adenosine triphosphate. Adenosine triphosphate, or ATP, is the form of energy directly utilizable by the cell.

When adequate insulin is not available, energy needs are met by catabolism of fats and protein. This conversion of fats and protein into a source of energy, termed *gluconeogenesis*, is a completely normal physiological phenomenon. Gluconeogenesis is a reserve mechanism which comes into play when energy demands exceed blood glucose and liver glycogen stores. When the body depends on fat and protein metabolism for most of its energy requirement, however, as occurs with diabetes, serious problems arise.

QUESTION 184   Diabetes mellitus is a disorder characterized by a deficiency of _____, a hormone produced by the _____, necessary for the cellular utilization of _____.

ANSWER   insulin ... pancreas ... glucose

QUESTION 185   When the body's supply of _____ is inadequate, blood _____ fails to enter the metabolic breakdown necessary for the synthesis of ATP, and _____ and _____ are called on to supply energy needs.

ANSWER   insulin ... glucose ... fats ... protein

QUESTION 186   Gluconeogenesis—that is, the conversion of _____ and _____ into _____ —is a normal physiological mechanism that provides _____ when stores of blood glucose and liver glycogen are diminished.

ANSWER   fats ... protein ... energy ... energy

ITEM 65 DIABETES—KETOACIDOSIS

Fats break down into glycerol and fatty acids, both of which can enter the glycolytic and tricarboxylic acid pathways for synthesis of ATP. When fatty acids are oxidized in excessive quantities, intermediate products, called *ketone bodies*, accumulate in blood. These ketone bodies include acetoacetic acid, β-hydroxybutyric acid, and acetone.

DIABETES MELLITUS

Uncontrolled diabetics oxidize such a large quantity of fatty acids for energy needs that ketone bodies build up excessively as a residue. An increased concentration of hydrogen ions results from this accumulation of acid ketone bodies in the blood, causing a form of metabolic acidosis known as *ketoacidosis*.

As hydrogen ion levels rise, part of the excess shifts into cells in exchange for potassium. Thus the ketoacidosis is compounded by hyperkalemia, which results both from intracellular release of potassium and from a lessened capacity of the kidneys to secrete potassium because of blockage of the secretory sites with hydrogen ions. A further complication of ketoacidosis is sodium depletion. Ketone bodies take sodium with them when they are excreted in their anionic form. Thus, sodium levels are depleted as acidosis develops from ketone bodies. Extremes of ketoacidosis can lower the pH of body fluids to 7.0 or below, thereby exceeding the lower limits for survival.

QUESTION 187
Ketone bodies appear in the blood when _____ _____ are excessively oxidized.

ANSWER
fatty acids

QUESTION 188
The specific ketone bodies which accumulate excessively in ketoacidosis include _____ acid, _____ acid, and _____.

ANSWER
acetoacetic . . . β-hydroxybutyric . . . acetone

QUESTION 189
Acidosis develops in the uncontrolled diabetic because of the accumulation of _____ _____ dissociated from _____ _____.

ANSWER
hydrogen ions . . . ketone bodies

QUESTION 190
When hyperkalemia accompanies ketoacidosis, the two causes of K⁺ build-up are (1) release of _____ potassium in exchange for _____ ions and (2) blockage by _____ ions of the available sites for renal secretion of _____.

ANSWER
intracellular . . . hydrogen . . . hydrogen . . . potassium

QUESTION 191
Sodium depletion may accompany ketoacidosis because acid _____ bodies take Na⁺ with them when they are excreted in their _____ form.

ANSWER
ketone . . . anionic

ITEM 66   COMPLICATIONS ASSOCIATED WITH DIABETES—ATHEROSCLEROSIS

A long-range complication of diabetes is *atherosclerosis*, which results from excessive levels of fats in circulating blood. Atherosclerosis leads to premature coronary artery disease and impaired circulation in the extremities. As the atherosclerotic condition advances, circulation is further impaired by the eventual development of *microangiopathy*, or thickening of the basement membrane of capillaries.

QUESTION 192

Atherosclerosis is a circulatory disorder characterized by high levels of _____ in circulating _____.

ANSWER

fats ... blood

QUESTION 193

Two principal sites of impaired circulation in atherosclerosis are the _____ and the arteries in _____.

ANSWER

coronary arteries ... extremities

QUESTION 194

Microangiopathy is a _____ of the basement membrane of _____.

ANSWER

thickening ... capillaries

ITEM 67   DIABETES—COMPENSATORY MECHANISMS

When diabetic acidosis develops, several compensatory mechanisms come into play to regulate pH of body fluids. Initially, excess hydrogen ions are reduced by combination with intracellular and extracellular buffers: protein/proteinate, bicarbonate/carbonic acid, disodium phosphate/monosodium phosphate, and so on. Sodium is reabsorbed in exchange for hydrogen and potassium ions across the tubular epithelium; urinary buffers absorb large amounts of hydrogen ions from tubular fluids; and ammonia synthesis is increased.

Through increased alveolar ventilation, hydrogen ions are removed from body fluids by increased expiration of carbon dioxide. The response of the respiratory system to metabolic acidosis, as a matter of fact, provides a very useful clinical aid in diagnosis of an acidotic diabetic. Increased rate and depth of respiration, particularly if the breath smells of acetone, are identifying symptoms in comatose patients who are brought into the emergency room. Blood tests that are positive for glucose and acid pH confirm the diagnosis.

The body combats diabetic acidosis through the reaction of the excess _____ _____ with the intracellular and extracellular buffers, including those of the _____, _____, and _____ systems.	QUESTION 195
hydrogen ions ... protein ... bicarbonate ... phosphate	ANSWER
The kidney compensates for excessive _____ _____ concentration by _____ exchange, reaction with the urinary _____, and synthesis of _____.	QUESTION 196
hydrogen ion ... sodium ... buffers ... ammonia	ANSWER
The respiratory system responds to diabetic acidosis by increasing the _____ and _____ of respiration in an effort to remove larger quantities of _____ from body fluids.	QUESTION 197
rate ... depth ... carbon dioxide	ANSWER
Diabetic acidosis can be identified in a comatose patient in the emergency room by the characteristic increase in _____ and the odor of _____.	QUESTION 198
respiration ... acetone	ANSWER
Blood tests for _____ and _____ are used to confirm the diagnosis of diabetic acidosis.	QUESTION 199
glucose ... pH	ANSWER

CLINICAL MANAGEMENT OF KETOACIDOSIS

ITEM 68

The patient approaching diabetic acidosis has probably reduced his food intake because of anorexia, nausea, and vomiting, so that intravenous electrolyte replacement may be necessary. Reduced water intake and possible vomiting, plus osmotic diuresis, frequently reduce extracellular volume to a level which dangerously depresses blood volume. If glomerular filtration rate responds to the decreased blood pressure to a point of anuria, acidosis may increase precipitously. Progression to a pH of 7.0 depresses the central nervous system, including the respiratory center. Breathing finally becomes shallow with respiratory depression, and the patient is in danger of diabetic coma and death.

Once ketoacidosis has been established from laboratory findings, treatment is aimed at restoring glucose utilization by administration of insulin. This procedure requires hospitalization so that frequent laboratory data can be used as an index for evaluation of insulin requirements.

QUESTION 200   Treatment of diabetic acidosis is aimed at restoring the metabolism of _____ by administration of _____.

ANSWER   glucose ... insulin

QUESTION 201   Extracellular fluid deficit may cause a drop in blood pressure and a decrease in _____ _____ which can lead to _____.

ANSWER   glomerular filtration ... anuria

QUESTION 202   With progression of pH to 7.0, acidosis depresses the _____ center of the central nervous system, so that breathing becomes _____, and the patient is in danger of diabetic _____ and death.

ANSWER   respiratory ... shallow ... coma

ITEM 69 KETOACIDOTIC SHOCK AND INSULIN SHOCK

If insulin is given in excessive dosage, the patient may have an adverse reaction. In both ketoacidotic shock and insulin shock, the patient may be comatose. In ketoacidosis, however, respirations are deep and rapid—unless respiratory depression has set in—and the breath smells of acetone. In insulin shock, respirations are shallow.

Treatment of ketoacidotic shock consists of administration of insulin and 50% glucose, which initiates and facilitates glucose entry into cells. Treatment of insulin shock consists of administration of glucose or glucose in orange juice.

When a diabetic patient is in shock, it is not immediately obvious whether the cause is lack of _____ or too much _____, insofar as in either case the patient may be _____.	QUESTION 203
<div align="center">insulin . . . insulin . . . comatose</div>	ANSWER
Deep and rapid respirations characterize the breathing of the comatose diabetic in *acidotic/insulin* (circle one) shock; shallow respirations, the patient in *acidotic/insulin* (circle one) shock.	QUESTION 204
<div align="center">acidotic . . . insulin</div>	ANSWER
The smell of acetone on the breath indicates that the comatose diabetic is in a state of _____ shock.	QUESTION 205
<div align="center">ketoacidotic</div>	ANSWER
The diabetic in _____ shock is given glucose in orange juice, while the patient in _____ shock receives insulin and 50% glucose.	QUESTION 206
<div align="center">insulin . . . acidotic</div>	ANSWER

GLUCOSE TOLERANCE TEST

ITEM 70

The glucose tolerance test is a means of detecting the effectiveness of a patient's insulin release in restoring blood glucose levels to normal after a test dose of glucose. Blood glucose levels are determined chemically before and at intervals after a large oral glucose load. A normal individual has a temporary rise in blood glucose, as illustrated in Figure 27, but the pancreas responds by excreting a larger quantity of insulin. Blood glucose levels return to normal in about three hours, and, in fact, show a hypoglycemic response.

The diabetic has a higher initial blood glucose concentration, shows a greater rise in blood glucose after ingestion of the glucose test dose, and still has an elevated level five hours later. The diabetic has this typical pattern of glucose tolerance response because the pancreas fails to secrete adequate amounts of insulin or the secreted insulin is chemically abnormal. Glucose is not channeled into cells as it is in normal individuals.

In hyperinsulinism, on the other hand, insulin release is exaggerated. The hyper-

280   CLINICAL CORRELATIONS

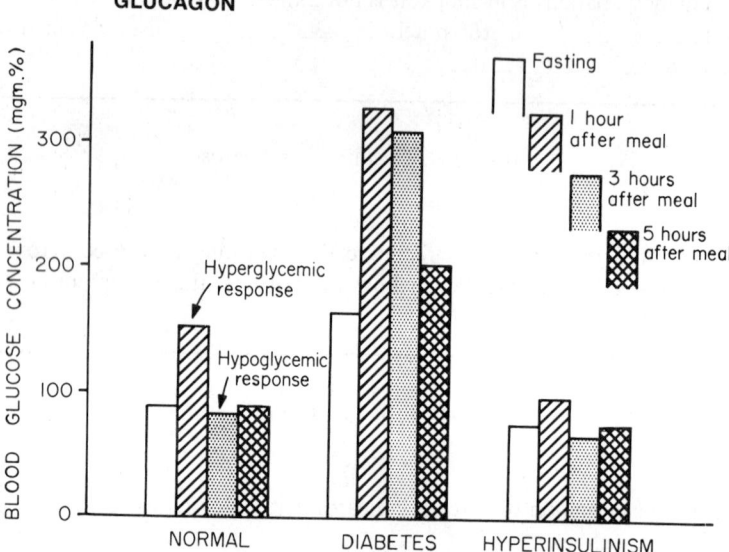

Figure 27. The effects of 50 gm. of glucose on blood glucose concentrations in normal subjects, diabetics, and hyperinsulinar patients. (From Guyton: *Function of the Human Body.* 4th Ed. Philadelphia: W. B. Saunders Company, 1974.)

glycemic response is not as pronounced, and blood glucose levels drop lower than normal during the hypoglycemic phase, as also can be seen in Figure 27.

QUESTION 207 The laboratory test which measures blood glucose levels before and after ingestion of a standard amount of glucose is called the _____ _____ test.

ANSWER glucose tolerance

QUESTION 208 In the laboratory test for diagnosing diabetes through measurement of blood glucose levels, a normal individual shows a temporary _____ in blood glucose value which returns to normal in about _____ hours, followed by a brief _____ response.

ANSWER rise (or increase) . . . three . . . hypoglycemic

QUESTION 209 The diabetic's blood glucose is _____ initially, reaches a still _____ level after ingestion of glucose, stays _____ longer, and shows no indication of _____ later on.

ANSWER higher . . . higher . . . elevated . . . hypoglycemia

DIABETES AND DISTURBANCE OF GLUCOSE HOMEOSTASIS

ITEM 71

Blood glucose levels remain elevated in the diabetic simply because of the failure of cellular utilization. Normal blood glucose values range from 75 to 100 mg./100 ml. The uncontrolled diabetic may have blood glucose levels two to three times normal. When hyperglycemia exceeds 180 mg./100 ml., the kidney is unable to reabsorb all the glucose filtered through the glomerular membrane, and the excess begins to be excreted in urine. Glucosuria noted in a routine urinalysis is indicative of possible diabetes.

Changes in body water and electrolytes in diabetes stem from the increased blood glucose level itself and from the accumulation of metabolic end-products of fat and protein utilization. Water balance is affected in the uncontrolled diabetic because of the urinary excretion of glucose.

The relative excess of solute in glomerular filtrate increases the quantity of water retained in renal tubules, so that osmotic diuresis causes a higher urinary volume. One of the first signs of diabetes is *polyuria* resulting from the high renal solute load. Another symptom of diabetes, *polydipsia*, or excessive thirst, results from stimulation of the thirst mechanism by the increased osmotic pressure of body fluids.

QUESTION 210
Absence of adequate _____ in the diabetic causes a failure of cellular utilization of blood _____, heightening the levels of _____ in blood.

ANSWER
insulin ... glucose ... glucose

QUESTION 211
Abnormally high _____ _____ is termed hyperglycemia; glucose appearing in _____ is termed glucosuria.

ANSWER
blood glucose ... urine

QUESTION 212
Problems in fluid balance in the diabetic stem from the urinary excretion of excess _____.

ANSWER
glucose

QUESTION 213
Excess glucose in blood is reflected in excess glucose in _____ _____, increasing the _____ pressure of the _____ _____.

ANSWER
glomerular filtrate ... osmotic ... glomerular filtrate

CLINICAL CORRELATIONS

QUESTION 214 The increase in urinary volume in diabetics is a result of _____ diuresis.

ANSWER osmotic

QUESTION 215 One of the first symptoms of diabetes is _____, which results from the high renal solute load.

ANSWER polyuria

ITEM 72 — INCIDENCE AND ONSET OF DIABETES

Diabetes is thought to be an inherited recessive trait and can appear in its clinically recognizable form anywhere from infancy to old age. Generally the earlier in life the disease becomes apparent, the more severe are its symptoms and the more erratic its course. Younger patients are prone to develop full-blown ketoacidosis with less provocation, and rigid insulin replacement is mandatory in such instances.

When onset occurs in later life, it is generally more easily controlled. "Maturity onset" diabetics are frequently maintained on dietary carbohydrate restriction and oral hypoglycemic preparations for years without incidents of extreme ketoacidosis.

Each patient presents an individual pattern, however, known best to the patient himself and to his physician. Whatever the diabetic state, practically all cases are more susceptible to diabetic acidosis during certain conditions of stress such as surgical trauma, infections, pregnancy, and the like. All such stresses call for increased insulin dosage.

Each hospital admission case, when diabetic acidosis is a possibility, requires careful history taking, contact with the patient's own physician whenever possible, and individualized therapy.

QUESTION 216 The tendency to develop diabetes is thought to be inherited, with the onset occurring at any age between _____ and _____.

ANSWER infancy . . . senescence (old age)

QUESTION 217 When diabetes develops in earlier life, its symptoms are likely to be more _____, with young patients more prone to develop _____ at less provocation.

ANSWER severe . . . ketoacidosis

When diabetes does not develop until maturity, the patient may need only restriction of _____ and _____, together with oral _____ preparations.

QUESTION 218

carbyhydrates ... sugar ... hypoglycemic

ANSWER

Diabetics are susceptible to acidosis during times of stress such as _____ trauma, _____, _____, and the like, at which times the need for insulin is increased.

QUESTION 219

surgical ... pregnancy ... infections

ANSWER

DIABETES: CASE HISTORY— DESCRIPTION OF THE CASE

ITEM 73

Mr. B., age 38, had been diagnosed as a borderline diabetic six months previously on a routine company physical examination. He was instructed as to dietary restrictions, was put on a 2000-calorie American Dietetics Association diet, and was taught how to test urine for sugar. In addition, an oral hypoglycemic medication was prescribed. His first reaction to the diagnosis and dietary restrictions was anger, because he had always prided himself on his excellent health. As a rapidly advancing executive in an insurance company, he was frequently involved in entertaining business associates and found adherence to the ADA diet troublesome.

For the past two weeks he had been so busy pursuing a large account that his eating pattern had become erratic. Not taking time to go to the company cafeteria for lunch, he began eating snacks. By the time he arrived home at night, he was so hungry that he was inclined to overeat. He completely discontinued the urine testing and took his medication only when he happened to remember it.

Mr. B. noticed that he was losing a little weight, but attributed it to his busy schedule. His excessive hunger and thirst would subside, he thought, when he was able to eat more regularly. He had noticed also some degree of polyuria, but he attributed this to the coffee and other drinks he had been consuming. Actually, the weight loss was due to fat and protein catabolism; the polyuria was brought on by the osmotic diuretic effect of large quantities of glucose in glomerular filtrate; the frequent hunger (polyphagia) was brought on by relative tissue starvation; and the polydipsia resulted from cellular dehydration. He was experiencing the four primary symptoms of diabetes: weight loss, polyuria, polyphagia, and polydipsia.

The polyuria indicative of Mr. B.'s diabetes stemmed from the osmotic force created in _____ _____ by the heavy solute concentration of _____.

QUESTION 220

glomerular filtrate ... glucose

ANSWER

284 CLINICAL CORRELATIONS

QUESTION 221 Mr. B.'s polydipsia was related to fluid loss through polyuria, which produced a degree of cellular _____.

ANSWER dehydration

QUESTION 222 Weight loss and polyphagia in diabetes are traceable to the same source, namely, tissue starvation from catabolism of _____ and _____.

ANSWER fats ... proteins

ITEM 74 — DIABETES: CASE HISTORY— SIGNS AND SYMPTOMS

During a sales meeting, Mr. B. developed a headache and vague stomach discomfort. He took two aspirins and continued the meeting under some stress. After the meeting, he ate a hurried lunch from the snack machine and later became nauseated and vomited. Finally, late in the afternoon he became drowsy and fell asleep at his desk. His secretary found him at closing time and awakened him with difficulty. She noticed that he had some shortness of breath and insisted that he call a doctor. Dr. M., fearing that the diabetes was out of control, made arrangements for Mr. B. to check into the hospital that night.

Admission notes described the patient as extremely lethargic and slow to respond. Breathing was loud and labored, and his breath smelled of acetone. The skin appeared dry. Blood pressure was 126/74; temperature, 100.4; pulse, 95; and respirations, 28/min.

The patient was placed on an NPO regimen, and an IV was started with normal saline. Blood was withdrawn to test for blood glucose, serum acetone and ketones, total CO_2 content, pH, sodium and potassium levels, BUN, and hematocrit. Intake and output records were begun, and monitoring of vital signs was carried out every hour. An EKG was started in order to monitor potassium levels.

QUESTION 223 On admission to the hospital, Mr. B.'s observable symptoms indicative of diabetic reaction included extreme _____, _____ breathing, and the smell of _____.

ANSWER lethargy ... labored ... acetone

QUESTION 224 Initial steps taken on the patient's admission included prohibition of any intake by _____; intravenous infusion with _____ _____ solution; and drawing blood for a series of tests to assess the status of his _____ condition.

ANSWER mouth ... normal saline ... diabetic

DIABETES: CASE HISTORY— LABORATORY FINDINGS

ITEM 75

Laboratory findings showed blood glucose 525 mg.% (normal is 75 to 100 mg.%), ketones 125 mg.% (normal is 0), and acetone 3+ (normal is 0). Total CO_2 content was 12 mM./L., compared with a normal of 23 to 27 for arterial blood. Serum sodium was elevated at 145 mEq./L., as was serum potassium at 5.2 mEq./L; pH was low at 7.1. BUN was high at 72 mg.%; hematocrit was high at 50%.

Mr. B. was hyperglycemic because of inadequate insulin secretion relative to his dietary intake and stress situation. The depressed total CO_2 content indicated a depletion of bicarbonate reserve secondary to ketoacidosis. The fact that Mr. B.'s pH was on the acidotic side indicated that his compensatory mechanisms were being exceeded by the ketoacidosis as well as by the lactic acidosis accompanying tissue anoxia and impaired circulation. The elevated serum sodium and hematocrit were due to dehydration resulting from the osmotic diuretic effect of glucosuria. Shift of intracellular potassium into extracellular fluid, as well as impaired secretion of potassium secondary to acidosis, accounted for Mr. B.'s being slightly hyperkalemic. The elevated BUN, finally, suggested some impairment of glomerular filtration, possibly the result of hypovolemia and decreased renal blood flow.

Laboratory values showed Mr. B. to be _____ and _____ due to inadequate insulin secretion.

QUESTION 225

---

hyperglycemic . . . ketoacidotic

ANSWER

Blood pH was low due to depletion of _____ reserve.

QUESTION 226

---

bicarbonate

ANSWER

Dehydration was indicated by the elevated serum _____ and the elevated _____.

QUESTION 227

---

sodium . . . hematocrit

ANSWER

Hyperkalemia was probably due to a shift of _____ potassium to _____ fluid and to impaired _____.

QUESTION 228

---

intracellular . . . extracellular . . . secretion

ANSWER

Impaired glomerular filtration is reflected in elevated _____.

QUESTION 229

BUN

ANSWER

ITEM 76 DIABETES: CASE HISTORY—TREATMENT

A few hours after treatment was initiated, Mr. B. was noticably more alert. He stated that the headache, nausea, and abdominal pain were subsiding.

Blood glucose determinations were made every four hours for the first day and insulin dosage determined accordingly. He received a total of 380 units of regular crystalline insulin intravenously during the initial 24-hour period. Intravenous administration of normal saline was continued at 1000 ml. q 8 hours × 2, followed by $D_5 \frac{1}{2} N/S \times 1$. If his blood pH had dropped sufficiently, sodium bicarbonate or sodium lactate might have been substituted for the saline.

During the first few hours of insulinization, Mr. B. was observed for hyperkalemia. This might have occurred because of a shift of intracellular potassium, with hydrogen ions, to the extracellular compartment. He was also observed for signs of bradycardia, weakness, flaccid paralysis, oliguria, and possible cardiac arrest.

For approximately 4 to 24 hours after initial treatment, there was a possibility that any hyperkalemia might rapidly shift to hypokalemia as potassium accompanied insulin-induced glucose entry into cells. Since the signs of hypokalemia are similar to those of hyperkalemia as well as paralytic ileus, it becomes necessary to rely on laboratory data and characteristic electrocardiographic findings to distinguish between these entities. Hyperkalemia is indicated by peaked T-waves, absent P-waves, and distorted QRS complexes. Hypokalemia is suspected when inverted or flattened T-waves and prolonged Q-T intervals are noted.

Since potassium levels had dropped to 3.6 mEq./L. at the end of 6 hours, oral potassium supplement was ordered.

QUESTION 230 Signs and symptoms of hyperkalemia include:

(1) _____

(2) _____

(3) _____ _____

(4) _____

ANSWER

(1) bradycardia

(2) weakness

(3) flaccid paralysis

(4) oliguria

QUESTION 231 Signs and symptoms of hypokalemia are similar to those of hyperkalemia and _____ _____ as well.

ANSWER paralytic ileus

Electrocardiographic findings in hyperkalemia include: **QUESTION 232**

(1) _____ T-waves

(2) _____ P-waves

(3) _____ QRS complexes

---

(1) peaked **ANSWER**

(2) absent

(3) distorted

Electrocardiographic findings in hypokalemia include: **QUESTION 233**

(1) _____ or _____ T-waves

(2) _____ Q-T intervals

---

(1) flattened ... inverted **ANSWER**

(2) prolonged

DIABETES: CASE HISTORY— FINAL TREATMENT AND DISMISSAL ITEM 77

Once insulin therapy was administered, Mr. B. was observed for signs of a hypoglycemic rebound. These include nervousness, weakness, sweating, headache, blurred vision, irritability, muscle incoordination, and apprehension. Since none of these symptoms appeared, he was considered out of immediate danger at the end of 48 hours. His IV was discontinued and he resumed oral fluids and his 2000-calorie AD diet.

During the remainder of his hospital stay, urine was tested with Clinitest and Acetest, and regular insulin was continued on a sliding scale. He attended the hospital's diabetic classes and was discharged at the end of a week, determined to keep his diabetes under control.

Signs and symptoms of hypoglycemia include: **QUESTION 234**

(1) _____

(2) _____

(3) _____

(4) _____

(5) _____

(Any five) **ANSWER**

(1) nervousness, (2) weakness, (3) sweating, (4) headache, (5) blurred vision, (6) irritability, (7) muscle incoordination, and (8) apprehension

CLINICAL CORRELATIONS

ITEM 78 EMPHYSEMA

Emphysema is a condition characterized by overinflation of alveoli and eventual rupture of alveolar walls, alveolar ducts, and respiratory bronchioles. These ruptures are accompanied by destruction of associated capillaries. With rupture of the walls and progressive enlargement of air spaces, functional respiratory membrane is reduced, and ventilation and perfusion are thereby diminished. Because of destruction of elastic tissue, the emphysematous patient experiences progressive impairment of the expiratory phase of respiration.

Eventually, the patient with emphysema develops a typical hyperinflated thorax. Adequate elastic recoil of the lungs is lost, and the diaphragm cannot relax to its normal dome-shape position on exhalation. Consequently, at the beginning of inspiration, the diaphragm starts contracting from a flattened baseline. The thoracic cage rises anteriorly, intercostal spaces enlarge, and the thorax assumes a greater anteroposterior diameter, described as a "barrel chest." Because of the overinflation, the patient utilizes accessory muscles of respiration to inspire, and the muscles of the neck and upper thorax stand out prominently on inspiration.

QUESTION 235
The complications of ventilation in the emphysematous patient affect more critically the *inspiratory/expiratory* (circle one) phase of breathing.

ANSWER
expiratory

QUESTION 236
The patient with emphysema has difficulty with inspiration because of the flattened position of the _____.

ANSWER
diaphragm

QUESTION 237
In emphysema, the difficulty in expiration stems from _____ of alveoli followed by _____ of alveolar walls and _____ of air spaces.

ANSWER
overinflation ... rupture ... enlargement

ITEM 79 SIGNS AND SYMPTOMS OF EMPHYSEMA

The major symptoms of emphysema are shortness of breath (due to ineffective expiration), dyspnea, hypoxia, and retention of carbon dioxide. Even with reduced activity and exertion, dyspnea becomes worse as the disease progresses. Respiration is rapid and shallow as build-up of carbon dioxide and hydrogen ions stimulates the respiratory center in the medulla. Unarrested, the hydrogen ion retention leads to respiratory acidosis and its sequelae.

Oxygenation is impaired because of the enlarged pulmonary dead air space which accompanies incomplete expiration. Some emphysematous patients may be pale and cyanotic as a result of reduced oxyhemoglobin in blood; others may have a ruddy complexion if hypoxemia has stimulated compensatory polycythemia. Wheezing and rales are present in proportion to the degree of trapping of secretions and inflammation.

The most common signs of emphysema are

(1) ──────────────

(2) ────────────── of breath

(3) ──────────────

(4) retention of ────────── ──────────

QUESTION 238

---

(1) dyspnea   (3) hypoxia

(2) shortness   (4) carbon dioxide

ANSWER

The emphysemic patient's complexion may be ────── because of reduced oxyhemoglobin, or it may be ────── because of compensatory polycythemia.

QUESTION 239

---

pale ... ruddy

ANSWER

Carbon dioxide build-up from incomplete expiration results (1) in retention of ────────── ions, leading to ────────── acidosis, and (2) in impaired ──────────.

QUESTION 240

---

hydrogen ... respiratory ... oxygenation

ANSWER

SYMPTOMS OF INCREASED pCO_2 WITH EMPHYSEMA

ITEM 80

Increased pCO_2 and decreased pO_2 in the cerebral circulation bring on mental changes. The patient may be lethargic and confused, and he may show impaired memory and poor judgment as well.

When the respiratory neurons in the medulla have been exposed to a longterm increase in pCO_2, they lose their normal sensitivity to the CO_2 stimulus to breathe. The respiratory center then relies, instead, on hypoxia as a stimulus through the carotid and aortic body reflexes. (While a slight increase in pCO_2 in a healthy individual stimulates the respiratory center, a drastic increase acts as a depressant, so much so that respirations may simply cease.)

The patient with a chronically elevated pCO_2 shows certain symptoms which are termed CO_2 narcosis. These symptoms include alterations in the sensorium, manifested as drowsiness, irritability, depression, and possible eventual coma. Depression of the respiratory center is evident as the patient hypoventilates and shows shallow respirations. Neuromuscular changes in CO_2 narcosis may include partial paralysis, tremors, and possible convulsions. Tachycardia and arrhythmias result from an initial cardiovascular compensation for hypoxia and from an eventual impairment of myocardial oxygenation.

QUESTION 241 Mental changes consequent upon increased pCO_2 in the cerebral circulation include _____, _____, and impaired _____.

ANSWER lethargy ... confusion ... memory (or judgment)

QUESTION 242 A *slight/great* (circle one) increase in pCO_2 stimulates the respiratory center; a *slight/great* (circle one) increase depresses this center.

ANSWER slight ... great

QUESTION 243 CO_2 narcosis is characterized by _____, _____, and _____.

ANSWER drowsiness ... irritability ... depression (any order)

QUESTION 244 Neuromuscular symptoms of CO_2 narcosis include partial _____, _____, and _____.

ANSWER paralysis ... tremors ... convulsions

ITEM 81 EMPHYSEMA AND BACTERIAL INFECTION

Along with the destruction of lung tissue which occurs in emphysema, cilia are also destroyed, and thus a normal means of raising pulmonary secretions is impaired. Bronchi tend to collapse erratically on expiration, an occurrence which traps air, mucus, and bacteria. Increased mucus formation due to chronic irritation, along with mucus retention, produces an environment favorable to bacterial growth and chronic infection. For this reason, emphysematous patients usually have an accompanying chronic bronchitis. Coughing is chronic, paroxysmal, and productive yet ineffective. The combination of emphysema and chronic bronchitis, along with asthma, is referred to collectively by the general term *chronic obstructive pulmonary disease*, abbreviated as COPD.

Because the destruction of lung tissue includes destruction of _____, an important means of removing secretions from the lungs is impaired in emphysema.	QUESTION 245
cilia	ANSWER
The inadequate expiration characteristic of emphysema causes the trapping of _____ and thereby creates a milieu favorable to growth of bacteria.	QUESTION 246
mucus	ANSWER
Chronic obstructive pulmonary disease (COPD) is a term which encompasses _____, chronic _____, and _____.	QUESTION 247
emphysema . . . bronchitis . . . asthma	ANSWER

EMPHYSEMA AND HEART FAILURE ITEM 82

As lung capillaries are progressively destroyed during the progress of emphysema, pulmonary hypertension develops. Increased pressure in the pulmonary circuit produces increased back pressure in the pulmonary artery and in the right ventricle. *Cor pulmonale*, the enlargement and failure of the right ventricle, can be caused by pulmonary hypertension. As right heart failure develops because the right ventricle is not able to propel blood effectively, central venous pressure increases and neck veins are distended. Eventually, high pressure in the capillary beds causes systemic edema and further reduction in tissue oxygenation.

In emphysema, back pressure of blood in the right ventricle results from destruction of lung _____.	QUESTION 248
capillaries	ANSWER

292 CLINICAL CORRELATIONS

QUESTION 249 With the progression of right heart failure, back pressure in the capillary beds causes both systemic _____ and reduced _____ of tissue.

ANSWER edema ... oxygenation

ITEM 83 EMPHYSEMA: CASE HISTORY— DESCRIPTION OF THE CASE

Mr. D., age 65, entered the hospital as a patient in the emergency department. A retired subway worker, he had fainted on a hot afternoon while mowing the grass. His wife, who was working in the yard with him, called a neighbor and together they moved Mr. D. into the shade. He revived quickly and reported that all of a sudden he had felt short of breath and dizzy. Mrs. D. called the family doctor, who arranged to meet them in the emergency department of the local hospital.

Hospital admission notes described Mr. D. as a thin, 65-year-old male in obvious respiratory distress. He was short of breath and dyspneic and appeared extremely anxious and irritable. Blood pressure was 138/84; temperature, 98.8; pulse, 104; respirations, 38/min. and shallow. Chest auscultation revealed bilateral rales in the lower lobes. Neck veins were distended and 1+ peripheral edema was noted in the ankles. Lips and nail beds appeared cyanotic, and the complexion was ruddy.

Mr. D. had several paroxysms of severe coughing when he attempted to talk. His wife explained that he had had a summer cold a few weeks ago and that he had been coughing a lot. Mr. D.'s doctor examined the patient and admitted him for tests. Blood was withdrawn and a urine specimen taken before his transfer from the emergency department.

QUESTION 250 Mr. D.'s visible symptoms on admission to the hospital were (1) _____ of _____; (2) _____; and (3) _____.

ANSWER shortness ... breath ... dyspnea ... anxiety (or irritability)

QUESTION 251 Mr. D.'s lips and nails were _____, and _____ were audible in the lower lobes.

ANSWER cyanotic ... rales

EMPHYSEMA: CASE HISTORY—INITIAL TREATMENT

ITEM 84

When Mr. D. arrived in his hospital room, oxygen was started at one liter per minute. The purposes of the low flow rate are to improve hemoglobin saturation and to counteract hypoxia without eliminating the hypoxic stimulus for breathing. Intermittent positive pressure breathing was initiated, and a β-adrenergic stimulator was administered to relax smooth muscle in bronchi. Mr. D. was given a bronchodilator to improve ventilation, an antibiotic to combat infection, and an expectorant to raise bronchial secretions. Intravenous infusion of $D_5 \frac{1}{2} N/S$ was utilized to combat the dehydration which may accompany a rapid respiratory rate. Visible signs of cyanosis and dyspnea were absent when the patient was checked before sleep.

Oxygen was administered to Mr. D. to counteract _____ and to enhance _____ saturation.

QUESTION 252

hypoxia ... hemoglobin

ANSWER

The rationale for the low flow rate of oxygen was that it would preserve the _____ stimulus for breathing.

QUESTION 253

hypoxic

ANSWER

The IV solution given Mr. D. was to combat _____.

QUESTION 254

dehydration

ANSWER

In addition to oxygen administration and IV infusion, other immediate clinical measures included giving Mr. D.

QUESTION 255

(1) a β-adrenergic stimulator to relax _____ _____ of bronchi

(2) a bronchodilator for _____

(3) an antibiotic for _____

(4) an _____ to raise secretions

---

(1) smooth muscle   (3) infection

(2) ventilation   (4) expectorant

ANSWER

ITEM 85 — EMPHYSEMA: CASE HISTORY— RADIOGRAPHIC FINDINGS

The next morning, Mr. D. was taken to the x-ray and nuclear medicine departments. Posterior to anterior and lateral radiographs revealed abnormal darkness of the lung field, due to loss of lung parenchyma. Widened intercostal spaces and a flattened diaphragm also appeared in the radiographs. A right anterior oblique radiograph revealed an enlarged right ventricle and pulmonary trunk. Ventilation and perfusion studies showed abnormalities in both parameters, which were explained by trapping of air and diminution of pulmonary circulation. Pulmonary function tests showed a reduced minute respiratory volume, reduced and slowed expiratory reserve volume, and an increased residual volume.

QUESTION 256 Radiographs revealed four significant aspects of Mr. D.'s conditions: (1) loss of lung _____ ; and (2) widened _____ spaces; (3) flattened _____ ; and (4) _____ right ventricle and pulmonary trunk.

ANSWER parenchyma ... intercostal ... diaphragm ... enlarged

QUESTION 257 Ventilation and perfusion abnormalities were due to trapping of _____ and poor pulmonary _____ .

ANSWER air ... circulation

ITEM 86 — EMPHYSEMA: CASE HISTORY— LABORATORY DATA

The initial laboratory analysis of the blood sample taken in the emergency department showed the following values: hematocrit elevated at 52%, red blood cell count elevated at 5.8 million per cu. mm., and hemoglobin high at 16.6 gm.%. Apparently Mr. D. had developed *polycythemia*, or abnormally high red blood cell production, as a mechanism to compensate for tissue hypoxia.

Analyses of arterial blood gases showed pCO_2 elevated at 120 mm. Hg (normal is 35 to 45 mm. Hg); pO_2 depressed at 46 mm. Hg (normal is 85 to 95 mm. Hg); O_2 saturation low at 54% (normal is 94 to 98%); total CO_2 content elevated at 36 mM./L. (normal is 23 to 27 mM./L.); arterial pH decreased at 7.28. A consideration of all these values indicated a respiratory acidosis resulting from carbon dioxide retention and a partially compensating metabolic alkalosis. The compensating alkalosis resulted from the kidney's retention of sodium bicarbonate in exchange for secreted hydrogen ions. Mr. D.'s IV solution was changed to Ringer's lactate to combat the respiratory acidosis.

Electrolyte values were normal except for potassium, which was slightly elevated at 5.2 mEq./L. Since the renal secretory machinery was overloaded with a large quantity of hydrogen ions, normal potassium secretion had been curtailed.

To counteract his tissue _____, Mr. D. had developed polycythemia, which is an excessively high production of _____ _____ _____.

QUESTION 258

hypoxia ... red blood cells

ANSWER

Laboratory values for Mr. D.'s pCO_2, pO_2, total CO_2 content, and low pH all pointed to a diagnosis of _____ acidosis.

QUESTION 259

respiratory

ANSWER

EMPHYSEMA: CASE HISTORY— ELECTROCARDIOGRAPHIC FINDINGS

ITEM 87

An EKG was ordered for evaluation of Mr. D.'s cardiac status and for monitoring of possible arrhythmias which might result from hyperkalemia. The EKG showed the typical hyperkalemic pattern: peaked T-waves, depressed S-T segment, widened QRS, and absent P-waves. To counteract the hyperkalemia, dietary potassium was restricted, and the patient was given a potassium exchange resin orally.

Vector cardiography confirmed the radiographic finding of right ventricular enlargement due to hypertrophy secondary to pulmonary hypertension. Mr. D. was diagnosed as having moderately advanced emphysema, bronchitis, and right heart failure.

The patient's EKG pattern displayed all the typical indications of _____.

QUESTION 260

hyperkalemia

ANSWER

Vector cardiography confirmed the radiographic evidence of the enlargement of the patient's _____ _____.

QUESTION 261

right ventricle

ANSWER

296 CLINICAL CORRELATIONS

QUESTION 262 From all the laboratory data, the final diagnosis was that Mr. D. had _____, _____, and _____ _____ failure.

ANSWER emphysema . . . bronchitis . . . right heart

ITEM 88 — EMPHYSEMA: CASE HISTORY—THERAPY AND DISMISSAL

Therapy continued throughout Mr. D.'s hospital stay and included use of antibiotics, expectorants, bronchodilators, and assisted positive pressure ventilation. He was taught exercises to improve respiratory function. Among the several exercises was that of exhaling through pursed lips to distribute air better and prevent bronchiolar collapse. He also learned forced exhaling by compression on the diaphragm.

Chest sounds were monitored frequently. With postural drainage, congestion and rales in the lower lobes became less distinct and finally disappeared.

The IV was discontinued after the third day of therapy, when it was felt that hydration was adequate to promote liquefaction of secretions.

As pulmonary ventilation improved, pCO_2 decreased and pO_2 increased toward normal. Arterial pH stabilized at 7.35, and total CO_2 content remained slightly elevated. Since lung damage is not reversible, the goal of treatment was to improve oxygenation to a level compatible with reduced activity. When Mr. D.'s lungs were clear and he was able to ambulate without dyspnea, his doctor discharged him after a ten-day stay in the hospital. The doctor outlined plans for Mr. D. to receive oxygen therapy and assisted ventilation at home.

QUESTION 263 Exhaling through _____ _____ and forced exhalation by compression of the _____ were taught as a means of preventing bronchiolar collapse.

ANSWER pursed lips . . . diaphragm

QUESTION 264 Since lung damage is irreparable, the rationale for treatment of emphysema is that _____ can be improved to support the reduced capacity of the pulmonary mechanism.

ANSWER oxygenation

BURNS—MULTIPLE CHARACTER OF THE PROBLEM

ITEM 89

Severe burns present multiple problems in fluid and electrolyte balance. Secretions of adrenal corticosteroids and antidiuretic hormone fluctuate from one phase of recovery to another, bringing on sodium and water balance fluctuations in their wake. Some fluid is lost in exudate and some is redistributed within the body in the form of edema. Sodium, needed to maintain blood volume, is initially retained. At a later stage, it is excreted in large quantities. Potassium released from damaged cells accumulates to dangerous levels in extracellular fluids during the first 24 to 48 hours, and then returns to intracellular fluid during the next few days. The general disruption of water and electrolyte balance, finally, has an adverse effect on urine output. Because of the generalized systemic response to trauma, essentially all aspects of fluid and electrolyte balance must be assayed carefully.

An additional complicating factor with severe burns is the fact that recovery proceeds in stages which require continual reevaluation. What may have been indicated therapy at one stage of recovery may be completely contraindicated at a later stage. An initial anuria, for instance, may reverse itself quickly to become diuresis. Hyperkalemia can revert to hypokalemia in a matter of hours.

Since severe burns constitute a multiphasic and continually changing problem, this condition challenges all the skill and judgment of the entire hospital burn team.

QUESTION 265

Fluid and electrolyte complications following severe burns present multiple problems: (1) Some fluid is lost in _____ and some is redistributed in _____ formation; (2) total body sodium is _____; (3) serum potassium levels are _____ during the first day or two, and then are _____ over the next few days.

ANSWER

(1) exudate . . . edema . . . (2) depleted . . . (3) increased . . . decreased

---

QUESTION 266

In response to the traumatic stress of severe burns, hormone secretions fluctuate, specifically _____ corticosteroids and _____ hormone.

ANSWER

adrenal . . . antidiuretic

QUESTION 267

The severely burned patient's fluid and electrolyte condition changes constantly during recovery; for example, an initial anuria may become _____; or it may happen that hyperkalemia can revert to _____ in a matter of hours.

ANSWER

diuresis . . . hypokalemia

ITEM 90

CLASSIFICATION OF BURNS ACCORDING TO DEPTH

According to depth of tissue destruction, burns are classified as first-, second-, or third-degree. A first-degree burn affects only the epidermis. Even though this type is moderately painful, recovery is spontaneous and usually requires no treatment except that designed to relieve hyperesthesia or increased sensitivity at the site. Localized vasodilation, with subsequent fluid leakage and edema formation, is minimal.

Second-degree burns, sometimes severely painful, extend to the dermal layer. Localized vasodilation and damaged capillaries in the area allow fluid to leak from plasma to interstitial space in larger quantities. Surface blisters form and break with leakage of exudate as edema fluids accumulate in the area. Broken skin surface imposes a danger of infection in the second-degree burn, so that sterile technique and antibacterial dressings are needed.

Third-degree burns extend to subcutaneous structures and result in extensive tissue destruction. This type is commonly without pain because nerve endings are destroyed, although pain may be felt from surrounding second-degree burns. The skin surface may be white and moist or charred and dry, later forming *eschar sloughs* as necrosed tissue is discarded. Deep burns are subject to infection and, if extensive, skin grafting is necessary, since the germinal epithelium is destroyed. Scarring and loss of function can be expected if tissue destruction is extreme.

QUESTION 268 In a first-degree burn, damage extends only to the _____, the skin appears _____, pain is _____, and treatment is designed to relieve pain.

ANSWER epidermis ... reddened ... moderate

QUESTION 269 A second-degree burn extends to the _____ layer, the skin appears _____, pain is _____, and recovery is sometimes complicated by _____.

ANSWER dermal ... blistered ... severe ... infection

QUESTION 270 A third-degree burn affects _____ structures, the skin appears _____ and _____ or charred and dry, pain is _____, and recovery is sometimes delayed by _____ and loss of function.

ANSWER subcutaneous ... white ... moist ... absent ... infection

DIAGNOSIS OF BURN DAMAGE AND DRUG THERAPY

ITEM 91

In the diagnosis of burns one must consider, in conjunction with the depth of localized damage, the percentage of total surface area involved. When burns involve from 20 to 50 per cent of the surface area of the body, life and death depend on knowledgeable management of fluid and electrolyte balance. Injury extending to greater than 50 per cent of the body offers increasingly less favorable prognosis for survival the greater the extent of involvement. Thus, second-degree burns over 30 per cent of the body, or third-degree burns over 10 per cent of the body, or burns involving vital organs, or those complicated by injury to the respiratory tract are all classified as critical.

Drug therapy for burns, in addition to relieving pain and combating infection, is designed to (1) maintain blood volume, (2) preserve kidney function, (3) reduce edema formation, and (4) stabilize electrolyte concentrations.

Burns must be diagnosed from two points of view: (1) according to the _____ of the burn and (2) according to the total _____ involved.

QUESTION 271

---

depth ... area

ANSWER

Burns that cover over _____ per cent of the total body area are considered critical; those covering over _____ per cent of the total body area may be fatal.

QUESTION 272

---

20 ... 50

ANSWER

The seriousness of the burn is critically increased if there is involvement of _____ organs or the _____ tract.

QUESTION 273

---

vital ... respiratory

ANSWER

Drug therapy for burns aims at

(1) maintaining _____ volume

(2) preserving _____ function

(3) reducing _____ formation

(4) stabilizing _____ concentrations

QUESTION 274

---

(1) blood

(2) kidney

(3) edema

(4) electrolyte

ANSWER

ITEM 92 — GENERAL FLUID SHIFTS FOLLOWING BURNS

The systemic response to second- and third-degree burns presents serious problems. The severely burned patient may go into hypovolemic shock as a consequence of fluid leaving the vascular system with edema formation. Loss of circulating blood volume disrupts the dynamics of glomerular filtration at a critical time, when kidney function is sorely needed.

Immediately after a severe burn, fluid begins to leak from serum to interstitial space through the damaged capillaries. Some of this fluid is lost from denuded areas of the skin and continues to be lost as long as a protective surface is not available to prevent exudation. Some of this fluid is trapped as edema and extends to non-burned areas of the body.

QUESTION 275 One of the earliest secondary dangers after a severe burn is _____ _____, as a result of fluid leaving the vascular system with _____ formation.

ANSWER hypovolemic shock ... edema

QUESTION 276 Secondary problems of burns may extend to the kidney in the form of reduced renal blood flow disrupting the dynamics of _____ _____ at a time when _____ function is sorely needed.

ANSWER glomerular filtration ... kidney

QUESTION 277 Initially, fluid leaks from _____ into _____ _____, some of the fluid being lost from denuded areas of the skin as _____ and some of the fluid being redistributed and trapped in the form of _____.

ANSWER serum ... interstitial space ... exudate ... edema

ITEM 93 — VASCULAR IMPLICATIONS IN FLUID SHIFTS FOLLOWING BURNS

Fluid will shift from serum into interstitial space as a result of the increased permeability accompanying vasodilation. Leakage of serum and blood from damaged blood vessels potentiates the overall shift of fluid into interstitial space. This potentiation occurs because the osmotic pressure in interstitial fluid is increased by leaked albumin, thus further promoting an osmotic water shift.

In areas where vascular continuity is destroyed, proteins and vascular volume are reduced, and the blood cells become more concentrated. If enough albumin is lost, hypoproteinemia then contributes to edema formation.

This particular fluid shift usually lasts about 48 hours after a burn is sustained. Intravenous therapy is indicated during this time to maintain blood volume and kidney function.

Shifts from serum to interstitial space occur because of increased capillary _____ and _____ blood vessels, resulting in a reduction in total blood _____, with the blood cells becoming more _____.

QUESTION 278

permeability . . . damaged . . . volume . . . concentrated

ANSWER

The particular pattern of fluid shifts following severe burns lasts about _____ hours, and intravenous therapy is needed to maintain blood _____ and _____ function.

QUESTION 279

48 . . . volume . . . kidney

ANSWER

ENDOCRINE RESPONSE TO BURNS

ITEM 94

During the first few hours following a burn, the endocrine system responds in several ways. The adrenal cortex secretes additional quantities of aldosterone and hydrocortisone as parts of the generalized stress reaction. Hormones are also released in defense against reduced blood volume and hemoconcentration. As aldosterone levels rise, the kidney reabsorbs additional sodium. Previous loss of fluid and electrolytes makes serum hyperosmolar. The pituitary gland secretes an increased quantity of antidiuretic hormone in response to this hyperosmolarity. With the increases in sodium reabsorption and antidiuretic hormone, water is retained. All these responses tend to restore blood volume but do so by retention of total available body water, since edema and exudation continue.

Decreased renal blood flow alters kidney function. The initial decrease in mean arterial blood pressure is reflected in a decreased renal blood flow and lowered glomerular filtration pressure. These filtration pressure changes reduce or halt altogether the urinary output, and complete anuria becomes a distinct threat.

CLINICAL CORRELATIONS

QUESTION 280 Endocrine response to burns includes (1) increase in aldosterone and hydrocortisone in response to _____, (2) increase in hormone release as a defense against reduced _____ volume, and (3) increase in antidiuretic hormone in response to the hyperosmolarity of _____.

ANSWER stress ... blood ... serum

QUESTION 281 The endocrine responses in cases of severe burns are directed toward restoring _____ volume by retention of _____ and _____.

ANSWER blood ... sodium ... water

QUESTION 282 Reduced blood volume affects the kidney in the following ways: (1) renal blood flow is _____, (2) glomerular filtration pressure is _____, and (3) urinary output is _____.

ANSWER decreased ... decreased ... decreased

ITEM 95 SODIUM LEVELS FOLLOWING BURNS

In addition to problems in fluid balance, problems in sodium and potassium levels occur within the first two days following severe burns. Sodium is affected early by the kidney's retention of this ion as a means of restoring blood volume. The overall picture, however, is usually one of sodium deficit because of other factors which override the aldosterone effect.

Since sodium is the main cation in extracellular fluid, sodium is displaced from the vascular system when it is trapped as edema fluid. Sodium deficit is aggravated when sodium-rich extracellular fluid leaks through damaged tissues and is lost as exudate. Serum sodium levels go down unless this ion is replaced by parenteral saline administration.

Along with the loss of sodium, loss of bicarbonate causes a base bicarbonate deficit and a tendency toward metabolic acidosis. Since acidosis is a decrease in the ratio of base to acid, the base deficit must be compensated for by adding bicarbonate to intravenous replacement solutions.

QUESTION 283 Even though sodium is retained by aldosterone following severe burns, there is an overall sodium loss because of the large quantity of this ion trapped as _____ fluid, and because of the quantity of this ion lost in _____.

ANSWER edema ... exudate

Intravenous saline infusion after burns is administered to compensate for _____ deficit.	QUESTION 284

sodium   ANSWER

Along with _____ depletion following burns, there is a corresponding loss of bicarbonate which sometimes upsets acid-base balance to the point of bringing on metabolic _____.	QUESTION 285

sodium ... acidosis   ANSWER

Acid-base balance can be restored in the acidotic patient by infusion of intravenous _____.	QUESTION 286

bicarbonate   ANSWER

POTASSIUM LEVELS FOLLOWING BURNS
ITEM 96

Serum potassium levels are up or down following burn damage, depending on the stage of recovery. Immediately following a burn, serum potassium levels are high as a result of cell damage itself. Disruption of cell membrane integrity in severe burns releases large quantities of intracellular potassium into extracellular fluids. The extent of serum potassium rise varies with the extent of cell damage.

In the early stage of recovery, when extracellular potassium levels are high, it is sometimes necessary to reduce the extent of hyperkalemia by forcing potassium entry into cells. This can be accomplished by administration of insulin. Insulin increases the transport of both glucose and potassium into cells, removing them from extracellular accumulation and thereby lowering potassium levels.

As recovery proceeds, serum potassium levels decline. Actually, hyperkalemia may become hypokalemia within a very few hours. This shift to low serum potassium occurs for two reasons. First, extracellular potassium shifts back into cells when cell membranes regain their restraining properties. Second, kidney function goes from a stage of oliguria to a stage of diuresis after a few days, so that any excess potassium is rapidly excreted.

Since potassium is concentrated in _____ fluids, disruption of cell membranes releases large amounts of potassium into _____ fluids.	QUESTION 287

intracellular ... extracellular   ANSWER

QUESTION 288
Dangerously high serum potassium levels can be lowered by administering _____, which forces potassium into the _____ compartment.

ANSWER: insulin ... intracellular

QUESTION 289
Serum potassium is high immediately after a severe burn because of cell _____; therefore, the extent of serum potassium increase is _____ to the extent of cell destruction.

ANSWER: destruction ... proportional

QUESTION 290
As recovery proceeds, serum potassium levels decline for two reasons: (1) intracellular potassium is no longer released when _____ membranes regain their restraining properties, and (2) kidney function goes from a stage of fluid _____ to a stage of _____.

ANSWER: cell ... retention ... diuresis

ITEM 97 FLUID SHIFTS DURING THE SECOND STAGE OF RECOVERY FROM BURNS

The second phase of recovery from burns involves a shift of interstitial fluid to blood as edema fluid reenters the vascular system. This phase begins 36 to 48 hours after the burn occurs and lasts for 24 to 48 hours. During this time, blood volume increases as edema fluid is mobilized. Hemodilution gives a decrease in laboratory electrolyte values and a decreased hematocrit. Increased blood volume increases renal blood flow and reverses oliguria to a stage of diuresis. As diuresis proceeds, total body sodium is decreased along with water. The tendency to metabolic acidosis continues with diuretic loss of sodium bicarbonate, although resumption of kidney function begins to provide compensatory adjustments to acid stress.

The need for intravenous expansion of blood volume which existed during the first day or two reverses to a point of endogenous circulatory overload, so that intravenous therapy is now contraindicated. Hypovolemia has become hypervolemia on the second or third day, as evidenced by increased urinary output. A failure of diuresis to occur at this point may be taken as evidence of possible renal damage. Kidneys can be damaged when red blood cell destruction releases excessive quantities of hemoglobin into plasma. Free hemoglobin in plasma causes diffuse renal damage by clogging the glomerular membrane and the renal tubules.

At about the second or third day following severe burns, capillary permeability is so improved that fluid shifts from _____ space to _____, and blood volume increases as _____ fluid returns to the _____ system.	QUESTION 291
interstitial ... blood ... edema ... circulatory	ANSWER
About two or three days following severe burns, laboratory values for hematocrit, hemoglobin, and electrolytes are likely to be _____ because of hemodilution.	QUESTION 292
reduced	ANSWER
Increased blood volume increases renal _____ _____ and reverses oliguria to a stage of _____.	QUESTION 293
blood flow ... diuresis	ANSWER
As urinary output increases, there is loss of total body _____ and _____, which may produce a tendency to continue metabolic _____ even though the kidney begins to resume excretion of hydrogen ions.	QUESTION 294
sodium ... bicarbonate ... acidosis	ANSWER
During the second phase of recovery from burns, excess intravenous therapy is contraindicated as hypovolemia reverses to _____, and the kidney responds by increasing _____ _____.	QUESTION 295
hypervolemia ... urinary output	ANSWER
Failure of the kidney to respond during the second phase of recovery from burns may be taken as evidence of possible kidney damage from a clogging of the nephrons with _____.	QUESTION 296
hemoglobin	ANSWER

ITEM 98: ELECTROLYTE SHIFTS DURING SECOND STAGE OF BURN RECOVERY

Sodium excretion may rise to twice normal during the diuretic phase of recovery from burns and begins to decrease slowly toward normal at about a week after the burn occurred. At this point, total body sodium levels can be replaced by oral intake or cautious hypertonic intravenous maintenance.

Hyperkalemia reverses to hypokalemia on about the fourth or fifth day as extracellular potassium moves back into cells. Since extremes of both excess and deficit of potassium must be avoided because of the threat to heart function, the doctor will order assays of serum potassium levels frequently.

In contrast to the first stage of recovery, when insulin is sometimes given to promote potassium entry into cells, potassium now enters cells spontaneously. This spontaneous entry, combined with a measure of urinary depletion due to diuresis, reduces serum levels to a point at which oral or intravenous potassium may have to be ordered.

Negative nitrogen balance may extend through the entire convalescent phase as a result of the generalized stress, tissue destruction, and tissue catabolism which accompany immobility. Protein continues to be lost as long as exudate seeps through burned areas of the skin. Serum protein levels may be maintained with plasma or whole blood during the first few days and by careful attention to a high protein oral intake as recovery proceeds.

QUESTION 297 During the diuretic phase following a burn, total body _____ is decreased and may have to be replaced by _____ or _____ dosage.

ANSWER sodium ... oral ... intravenous

QUESTION 298 Potassium levels fluctuate from _____ during the early stage of recovery from burns to _____ on the fourth or fifth day.

ANSWER hyperkalemia ... hypokalemia

QUESTION 299 During the early stage of recovery, extracellular potassium levels are _____ because of release of potassium from _____ fluid.

ANSWER high ... intracellular

QUESTION 300 During the second stage of recovery, potassium levels are _____ because of reentry of potassium into _____ and because of increased kidney excretion.

ANSWER low ... cells

Both high and low levels of serum potassium must be avoided because of danger to _____ function.	QUESTION 301
heart	ANSWER
During the first stage of recovery, serum potassium levels can be lowered by giving _____, whereas in the second stage of recovery, hypokalemia can be controlled by oral or intravenous administration of _____.	QUESTION 302
insulin . . . potassium	ANSWER
Generalized stress, tissue destruction, and tissue catabolism affect protein stores and result in negative _____ balance.	QUESTION 303
nitrogen	ANSWER
Serum protein levels may be maintained by giving plasma or whole blood, followed by a high _____ intake.	QUESTION 304
protein	ANSWER

BURNS: CASE HISTORY— IMMEDIATE EMERGENCY TREATMENT

ITEM 99

Mrs. N., age 35, was severely burned when her home caught fire. She was asleep in a second-floor bedroom, and by the time she awakened, the stairway was in flames. When firemen rescued her through a window, it was apparent that she had sustained extensive burns.

Large areas of her skin were charred, and she was almost unconscious. An ambulance took her to the emergency hospital, where the burn team had been alerted to her arrival. Mrs. N. was given a dose of morphine to relieve pain, and initial procedures were begun.

The burn cart contained an IV tray with needles and syringes, a cut-down tray, sterile gowns and gloves, sterile dressings, lactated Ringer's solution, and 5% dextrose in water. These articles on a burn cart are periodically checked for sterility and replaced as necessary, so that no time is lost at the patient's bedside.

Blood was withdrawn from the left median cubital vein for the initial laboratory work-up. A sample was sent to the lab for blood typing.

308 CLINICAL CORRELATIONS

QUESTION 305
A hospital burn cart routinely includes, in addition to sterile instruments and dressings, two solutions: (1) _____ _____ solution and (2) _____ _____ in _____.

ANSWER
lactated Ringer's . . . 5% dextrose . . . water

ITEM 100 BURNS: CASE HISTORY— FIRST DAY

As the doctor carefully removed loose bits of clothing, the nurse recorded details of the accident and noted the general appearance of the patient on arrival. Large areas of the lower extremities and lower right quadrant appeared blistered and moist. Other areas were charred and dry. Unburned areas of the skin were pale and felt cold to the touch. A heart rate of 110 beats per minute indicated a tachycardia induced as a reflex compensation to maintain blood pressure. Recorded blood pressure was 110/80. Mrs. N. was disoriented and restless and appeared to be in shock.

The doctor prepared for immediate intravenous infusion of dextran as a temporary blood volume replacement, and 500 ml. were administered in the emergency department. Blood type was later available from the lab, and the infusion was changed to 500 ml. whole blood replacement.

Mrs. N.'s burns were cleaned by removal of loose epidermis, and Sulfamylon was applied. It was estimated that second-degree burns covered 28 per cent of the body and third-degree burns covered 18 per cent. Morphine sulfate was continued for relief of pain. An indwelling urinary catheter was installed and Mrs. N. was assigned to a room.

Initial orders included fluid balance recordings, hourly recordings of vital signs, and oral electrolyte solution as needed for thirst.

QUESTION 306
On arrival, Mrs. N.'s general condition was recorded: she had some first-degree burn, where the skin was _____; some second-degree burn, where the skin appeared _____ and _____; and some third-degree burn, where the skin was _____ and _____.

ANSWER
reddened . . . blistered . . . moist . . . charred . . . dry

QUESTION 307
Mrs. N. showed some symptoms of shock. These were disorientation and tachycardia as well as _____ blood pressure and unburned areas of the skin which were _____ and _____.

ANSWER
low . . . pale . . . cold

Initial burn shock is routinely handled by maintenance of adequate _____ volume by _____ infusion.	QUESTION 308
blood . . . intravenous	ANSWER
Mrs. N. was given _____ until her blood type was available from the laboratory.	QUESTION 309
dextran	ANSWER
Loose epidermis was removed and a wet dressing of _____ was applied; pain was relieved with _____ sulfate; as a means of detecting early acute renal shutdown due to low blood pressure, an indwelling _____ was installed.	QUESTION 310
Sulfamylon . . . morphine . . . catheter	ANSWER
Initial orders included recordings of _____ balance, hourly reports on vital signs, and administration of oral _____ solution.	QUESTION 311
fluid . . . electrolyte	ANSWER

BURNS: CASE HISTORY— SECOND DAY
ITEM 101

Intravenous therapy was continued for 36 hours to prevent shock. The infusion was patterned after the Evans formula, a standard used in many hospitals. This formula suggests 1 ml. colloid per kilogram of body weight per percentage of body surface burned.

Mrs. N. weighed 62 kg. and approximately 46 per cent of her body was affected. This equaled 1 ml. colloid × 62 kg. × 46 per cent burned area, or 2850 ml. colloid. The Evans formula suggests an additional equal volume of balanced electrolyte solution and 1500 ml. of 5% dextrose in water for hydration. Thus the intravenous solution prescribed for Mrs. N. included:
   2850 ml. colloid, or whole blood,
   2850 ml. lactated Ringer's solution, and
   1500 ml. 5% dextrose in water.

Orders were changed on the second day, after the initial danger of hypovolemic shock had passed, to read 2 liters 5% dextrose in water for hydration and 3 liters of 3:1 lactated

310 CLINICAL CORRELATIONS

Ringer's solution and colloid. Recorded input included the above infusions and 850 ml. oral electrolyte solution in the form of chipped ice, as the patient needed it for thirst.

Urinary output for the first 24 hours was 850 ml.; for the second 24 hours, 1475 ml. Since it appeared that renal function was established, the catheter was removed.

QUESTION 312

The Evans formula was followed to prevent _____.

ANSWER

shock

QUESTION 313

The standard formula for the Evans solution is 1 ml. _____ per kilogram of body weight times the percentage of burned body surface, to which is added an equal volume of _____ _____ solution and 1500 ml. _____ _____ in water.

ANSWER

colloid ... balanced electrolyte (or lactated Ringer's) ... 5% dextrose

QUESTION 314

On the second day, Mrs. N.'s intravenous replacement was altered to read two liters of 5% dextrose in water for _____, and 3 liters of 3:1 _____ _____ solution and _____.

ANSWER

hydration ... lactated Ringer's ... colloid

QUESTION 315

Mrs. N. was given oral electrolyte solution in the form of _____ _____ for thirst.

ANSWER

chipped ice

QUESTION 316

Urinary output for the first day or two following severe burn is likely to be _____ until blood volume stabilizes.

ANSWER

decreased

ITEM 102 BURNS: CASE HISTORY—LABORATORY WORK-UP

The staff was constantly alert for signs indicating the adequacy of fluid therapy. They watched for evidence of insufficient fluid intake, which included decreased urinary output, collapsed veins, restlessness and disorientation, hypotension, and tachycardia.

Signs of excess intravenous infusion are dyspnea, venous distention, moist rales, possible decreased hematocrit, and increased blood pressure. None of these symptoms appeared during the first 48 hours, and blood pressure increased to 130/90. Intravenous therapy was discontinued after 48 hours, but oral intake/urinary output volume studies were continued.

Initial laboratory findings at the end of 24 hours showed hematocrit, 32; hemoglobin, 10 grams; BUN, 20 mg./100 ml.; serum sodium, 146 mEq./L.; and serum potassium, 7.5 mEq./L. An injection of insulin reduced potassium levels to 5.0 mEq./L. Serum sodium and potassium were relatively unchanged on the next day, and it was felt that these electrolytes had stabilized.

After the danger of post-burn shock is passed, _____ therapy is evaluated on an ongoing basis.

QUESTION 317

---

intravenous

ANSWER

Inadequate fluid infusion is to be avoided because of _____; overextended fluid infusion is to be avoided because of circulatory _____; the best indication of adequate fluid therapy is _____ and _____ balance.

QUESTION 318

---

shock ... overload ... intake ... output

ANSWER

Symptoms of inadequate fluid therapy include restlessness, disorientation, and tachycardia, as well as hypotension, _____ urinary output, and collapsed _____.

QUESTION 319

---

decreased ... veins

ANSWER

Symptoms of overextended fluid therapy include

(1) _____

(2) venous _____

(3) moist _____

(4) increased _____ _____

QUESTION 320

---

(1) dyspnea

(2) distention

(3) rales

(4) blood pressure

ANSWER

312 CLINICAL CORRELATIONS

QUESTION 321   Since lab reports at the end of the first day showed Mrs. N.'s potassium levels were _____, serum potassium was _____ by giving _____.

ANSWER   high ... reduced ... insulin

QUESTION 322   Hematocrit was slightly decreased, indicating _____.

ANSWER   hemodilution

ITEM 103   BURNS: CASE HISTORY— THIRD DAY

Mrs. N. entered the hospital on Monday. By Wednesday night she showed the normal progress in recovery. Blood volume stabilized as edema fluid was reabsorbed into the vascular system. Capillaries had healed enough that fluid was no longer lost into interstitial space. With improved capillary dynamics and increased blood volume, renal blood flow increased, glomerular filtration rate increased, and the diuretic phase of fluid adjustment began.

The patient was continually observed for signs of excess circulatory overload which could lead to a fatal pulmonary edema. Parenteral fluids had been discontinued after 48 hours in anticipation of this typical pattern of fluid shifts, and intake was limited to oral electrolyte solution. A high-protein, high-calorie diet was started on the third day to supply adequate nutrients for the tissue-rebuilding process ahead.

QUESTION 323   Fluid reenters the _____ system during the second phase of recovery when capillaries are sufficiently healed to maintain _____ volume.

ANSWER   circulatory ... blood

QUESTION 324   With improved capillary dynamics, fluid goes from _____ _____ to _____ as _____ fluids are mobilized.

ANSWER   interstitial space ... blood ... edema

Increased blood volume increases _____ _____, and the patient goes from a state of oliguria, during the first two days, to a state of _____ lasting over the next 24 to 72 hours.

QUESTION 325

urinary output ... diuresis (or polyuria)

ANSWER

When fluid shifts from _____ space to serum, _____ therapy is contraindicated.

QUESTION 326

interstitial ... intravenous

ANSWER

After parenteral fluids are discontinued, fluid intake is limited to oral _____ solution; diet is high in _____ and _____ intake for tissue rebuilding.

QUESTION 327

electrolyte ... protein ... calorie

ANSWER

BURNS: CASE HISTORY—POTASSIUM THERAPY

ITEM 104

In burn patients such as Mrs. N., it is possible for blood potassium levels to reverse from a hyperkalemia, present during the first day or two, to a frank hypokalemia, occurring as potassium reenters cells or is excreted. Hypokalemia is recognized as a problem when extracellular fluid levels drop below 3.5 mEq./L.

The medical staff watched Mrs. N. for the clinical signs of low potassium balance. These include (a) generalized weakness and muscle flaccidity, (b) hyporeflexia, (c) weak pulse and faint heart sounds, (d) falling blood pressure, (e) vomiting, and (f) shallow respirations. Although none of these symptoms were apparent, Mrs. N.'s laboratory work-up showed serum potassium was at 3.7 mEq./L. Potassium was added to the oral electrolyte solution, and fruit juices were given to ensure an adequate intake of this ion. It was felt that kidney function was sufficiently restored, so that the danger of retention of excess potassium was no problem.

Potassium levels, during the second stage of recovery, shift from _____ to _____ because of reentry of potassium into _____ and because of _____.

QUESTION 328

hyperkalemia ... hypokalemia ... cells ... diuresis

ANSWER

314 CLINICAL CORRELATIONS

QUESTION 329 Clinical symptoms of hypokalemia, in addition to hyporeflexia, vomiting, and generalized weakness, are

(1) muscle _____

(2) _____ pulse and _____ heart sounds

(3) _____ blood pressure

(4) _____ respirations

ANSWER
(1) flaccidity
(2) weak . . . faint
(3) falling
(4) shallow

QUESTION 330 Although Mrs. N. did not manifest the symptoms of hypokalemia, her serum potassium levels were slightly low, so potassium was added to the oral _____ solutions, and she was fed _____ _____ .

ANSWER electrolyte . . . fruit juices

QUESTION 331 Mrs. N.'s kidney function had sufficiently stabilized by the third day that there was little danger of _____ of overextended potassium replacement.

ANSWER retention

ITEM 105 BURNS: CASE HISTORY— FINAL TREATMENT

A daily fluid balance summary showed that the patient had experienced the typical response of the vascular system to burns. A fluid accumulation phase began immediately after the injury and continued for 24 to 48 hours. During this time, edema fluids accumulated and urinary volume diminished. In more severe cases, edema fluid can accumulate so diffusely that its volume approaches that of a normal serum volume.

The fluid remobilization, or diuretic, phase began the second day and continued for 24 to 72 hours. Mrs. N.'s chart showed that she was following the usual pattern. Apparently no damage had been imposed on the kidneys.

The fluid intake/output summary included the following:

24-hr period	Intake in ml.	Output in ml.	Total body water balance
1st	500 Dextran 500 Whole blood 2850 Colloid 2850 Lactated Ringer's	850 Urinary output	
	6700 Total	850 Total	Positive
2nd	2000 Dextrose 3000 Electrolyte & colloid 850 Oral electrolyte	1675 Urinary output	
	5850 Total	1675 Total	Positive
3rd	550 Orange juice 225 Grapefruit juice 470 Oral electrolyte	5220 Urinary output	
	1245 Total	5220 Total	Negative
4th	580 Orange juice 365 Grapefruit juice 200 Oral electrolyte	5745 Urinary output	
	1145 Total	5745 Total	Negative

After the fourth day, Mrs. N.'s edema was negligible, and plain water was allowed as desired. Fluid intake/output recordings were discontinued on the seventh day.

Mrs. N.'s burns were debrided as needed, and Sulfamylon medications were continued throughout the healing process. Auto-grafting was begun on the twelfth day and was completed on the thirty-fifth day. Infection was kept at a minimum throughout the course of recovery.

Physical therapy was instituted and successfully restored function to the severely burned lower extremities. Rehabilitation proceeded uneventfully and Mrs. N. was discharged from the hospital on the thirty-eighth day.

QUESTION 332

Fluid balance following a severe burn follows a definite course of events: during the first two days, fluid shifts from _____ to _____ _____; during the second stage, fluid goes from _____ _____ to _____ as _____ fluids reenter the vascular system.

ANSWER

serum . . . interstitial space . . . interstitial space . . . serum . . . edema

QUESTION 333

Total body water goes from a state of fluid _____ during the early phase to a state of fluid _____ during the diuretic phase.

ANSWER

accumulation . . . deficit

QUESTION 334

During the first two days, fluid intake/output charts showed a _____ total body water balance when fluid was retained; after the second day, fluid charts showed a _____ total body water balance as edema fluids were excreted.

ANSWER

positive . . . negative

316 CLINICAL CORRELATIONS

QUESTION 335
Plain water was allowed as desired when fluid balance was _____ and kidney function was established.

ANSWER
negative

QUESTION 336
Infection was treated with _____ during the skin grafting, and _____ _____ restored function to the injured lower extremities.

ANSWER
antibiotics ... physical therapy

SUMMARY OF UNIT FIVE

Fluid and electrolyte problems arising secondary to disease, injuries, and malfunctions often pose a greater threat to a patient than the original clinical condition.

Postoperative Management of Fluids and Electrolytes

The combined responses of the pituitary and adrenal glands mediate the endocrine response to the tissue destruction involved in surgery. Hydrocortisone modifies metabolism for utilization of fat and protein for tissue repair. Aldosterone and other hormones promote water retention and thus maintain circulatory dynamics to protect vital structures against variable blood volume. Sodium retention and potassium depletion are the principal electrolyte changes brought on by increased aldosterone.

Following surgery, cells hydrolyze protein for utilization as energy. This protein catabolism is revealed by increased nitrogen excretion, mostly in the form of urea. When renal insufficiency threatens, protein is catabolized but not excreted.

As solute in extracellular fluids rises, fluid moves by osmosis from the intracellular to the extracellular compartment. About a week following surgery, the adrenal reaction subsides, gluconeogenesis slows, the system returns to carbohydrate metabolism as a source of energy, and a positive nitrogen balance is restored.

The vital organs which the body reflexively protects following surgery are the heart, brain, lungs, and kidneys. Constriction or dilation of arterioles distributes blood volume in such a way that non-vital areas receive less in deference to the needs of vital areas.

During the post-surgical period, lungs and kidneys share the task of maintaining water and acid-base balance; but maintaining electrolyte composition is, almost entirely, the work of the kidneys alone. In the matter of acid-base balance, the lungs are crucial because of the relatively enormous numbers of hydrogen ions removed daily in normal alveolar ventilation. Any serious breakdown of respiration leads to a dangerous acidosis.

Imbalance Due to Heart Failure

Congestive heart failure is the inability of the heart to pump enough blood to meet normal circulatory demands. If the left ventricle fails to propel enough blood, increased venous return stretches the heart muscle to the point of weakening contractile strength; back pressure builds up, causing a rise of pulmonary capillary pressure; fluid leaks into the tissue spaces of the lungs; and pulmonary edema results.

When congestive failure stems from failure of the right side of the heart, certain reflexes go into operation to return blood pressure to normal. These reflexes bring about a reduction of glomerular filtration, so that the kidneys retain fluid for the maintenance of blood volume. Additional blood volume, coupled with inadequate pumping, heightens hydrostatic pressure of all capillary beds. Fluid leaks into tissue space throughout the body, causing a generalized edema, particularly apparent in the lower extremities.

To bring about the fluid retention necessary to restore blood volume, the adrenal cortex releases aldosterone, which promotes sodium retention.

Treatment of heart failure includes administration of digitalis, bed rest, and restriction of salt intake. Digitalis effects calcium influx into contractile proteins to strengthen contractility of heart fibers. Bed rest reduces work load and encourages diuresis. Restricted salt intake reduces fluid retention.

Imbalance Due to Liver Disease

In liver disease, the connective tissue replacing destroyed liver cells clogs the lumina of the small branches of the portal vein, increasing back pressure of the blood in the veins of the abdomen. Fluid leaks from these abdominal veins and accumulates in the peritoneal cavity, bringing on ascites.

The accumulation of ascites fluid lowers serum protein levels because this fluid has a high content of protein absorbed from the intestine. Hypoproteinemia is aggravated in liver disease because of the liver's diminished ability to synthesize albumin, fibrinogen, and alpha and beta globulins.

For reasons unknown, increased circulating levels of antidiuretic hormone and aldosterone accompany liver disease. The excess of antidiuretic hormone increases reabsorption of water, reduces urinary output, and brings on generalized edema. Excess aldosterone increases sodium retention, reduces potassium levels, and creates a fluid environment conducive to cardiac arrhythmias.

Imbalance Due to Renal Insufficiency

Acute renal failure, or loss of kidney function, results from some primary trauma such as injury, heart failure, or infection, and it usually subsides when the original cause is removed. Chronic renal failure, on the other hand, is a gradual degenerative disease involving the progressive destruction of nephrons. Both forms manifest themselves in edema, acidosis, hyperkalemia, elevated non-protein nitrogen, and retention of urea.

Renal failure, acute or chronic, may result in either excessive retention or excessive excretion of both sodium and potassium. Accurate, repeated laboratory analyses will dictate whether excretion or conservation of salts is needed to control serum electrolyte balance.

As renal failure grossly reduces ammonia synthesis and eventuates in hydrogen ion build-up, the patient may become acidotic. Treatment will aim at the maintenance of acid-base balance.

When the patient with renal failure becomes oliguric, both water and nitrogenous metabolites are retained and bring on edema and uremia. To counteract this condition, protein and water intake must be regulated carefully.

Secondary effects of kidney disease include greater susceptibility to infection, bleeding tendency, hypertension, and imposition of an additional work load on the heart.

Standard temporary expedients for preserving optimum concentrations of extracellular fluids are hemodialysis and peritoneal dialysis. These measures remove toxic substances and conserve needed electrolytes. The specific composition of the dialysis solution in each case will determine which electrolytes are removed and which are restored.

Imbalance Due to Diabetes

Because of inadequate supply of insulin to promote cellular utilization of glucose, the diabetic metabolizes fats and proteins to meet energy needs. Acid ketone bodies accumulate in the blood, hydrogen ion concentration increases, sodium depletion accompanies excretion of part of the hydrogen ion excess, and the pH of body fluids moves in the direction of the lower limit for survival.

The glucose tolerance test measures blood glucose levels before and after oral ingestion of glucose. In the diabetic, blood glucose levels are significantly higher and remain elevated for as long as five hours after the ingestion of glucose. Detection of glucose in the urine also indicates diabetes. Polyuria and polydipsia are symptomatic of diabetes.

In cases of diabetic acidosis, the body attempts to regulate the pH of body fluids through the various buffer systems, renal excretion, and increased alveolar ventilation. Clinical treatment of diabetic ketoacidosis aims at restoring glucose utilization by administration of insulin.

Imbalance Due to Emphysema

Emphysema results from rupture of alveolar walls, ducts, and respiratory bronchioles. Associated capillaries are destroyed, air spaces enlarge, functional respiratory membrane is reduced, and the expiratory phase of respiration is progressively impaired.

Signs and symptoms of emphysema include shortness of breath, dyspnea, hypoxia, lethargy, and retention of carbon dioxide. Carbon dioxide and hydrogen ion build-up lead to respiratory acidosis. Oxygenation is inadequate, and chronically elevated pCO_2 may produce convulsions, tachycardia, and arrhythmias. Increased pressure in the pulmonary circuit creates back pressure in the pulmonary artery and right ventricle. Right heart failure, systemic edema, and poor tissue oxygenation ensue.

Since damage to lung tissue is irreparable, the aim of treatment is to improve oxygenation at a level compatible with the lungs' restricted capacity and reduced activity. Antibiotics, bronchodilators, and assisted positive pressure breathing are among the means of achieving the goals of therapy.

Imbalance Due to Burns

First-degree burns extend to the epidermis; second-degree, to the dermal layer; and third-degree, to subcutaneous structures. First-degree burns are critical if they extend over 50 per cent of the body surface; second-degree, if over 30 per cent; and third-degree, if over 10 per cent. In all these cases, drug therapy helps relieve pain, combat infection, sustain blood volume, preserve kidney function, reduce edema formation, and stabilize electrolyte concentrations.

The fluid and electrolyte problems immediately following severe burns include (1) edema formation, which begins within minutes and extends to about 48 hours after a burn is sustained, with serum loss of fluid and electrolytes exceeding loss of protein; (2) high protein fluid loss from denuded areas of the skin, continuing as long as these areas remain exposed; (3) total body sodium deficit, as high-sodium-content extracellular fluids are lost as exudate; and (4) hyperkalemia owing to cellular destruction releasing intracellular potassium into extracellular fluids. During this first phase following severe burns, intravenous expansion of blood volume is indicated.

After the first 36 to 48 hours following burns, a second phase of recovery occurs in which fluid returns to the blood volume. Events in the second phase are the reverse of those in the first phase: blood volume increases, with a danger of circulatory overload; oliguria is supplanted by diuresis; total body sodium is excreted with water, and metabolic acidosis becomes a distinct threat; potassium reenters cells spontaneously and depletes serum potassium levels. Intravenous or oral intake of sodium and potassium may be needed to maintain electrolyte balance.

The endocrine response to surgical tissue trauma is mediated primarily by the _____ gland and the _____ glands. QUESTION 337

---

pituitary ... adrenal ANSWER

Endocrine responses following surgery are aimed at maintaining _____ _____ to protect _____ _____. QUESTION 338

---

blood volume ... vital organs (or brain, heart) ANSWER

Increased aldosterone secretion following surgery can lead to _____ retention and _____ depletion. QUESTION 339

---

sodium ... potassium ANSWER

Following surgery, cells hydrolize protein for energy; when protein is not excreted, _____ _____ is suspected. QUESTION 340

---

renal insufficiency ANSWER

Respiratory function following surgery is critical for regulating acid-base balance because of the large quantity of _____ _____ removed by normal breathing. QUESTION 341

---

hydrogen ions ANSWER

Maintaining electrolyte balance following surgery is done almost exclusively by the _____. QUESTION 342

---

kidneys ANSWER

CLINICAL CORRELATIONS

QUESTION 343 The effect of digitalis is to bring _____ into contractile proteins to _____ contraction of the heart.

ANSWER calcium ... strengthen

QUESTION 344 Obstruction of blood flow through the liver increases back pressure in the _____ vein and in the veins of the _____, bringing on an accumulation of _____ fluid.

ANSWER portal ... intestines ... ascites

QUESTION 345 Excess circulating levels of antidiuretic hormone accompanying liver disease promote _____ of water and generalized _____.

ANSWER reabsorption ... edema

QUESTION 346 Excess circulating levels of aldosterone accompanying liver disease increase _____ retention, reduce _____ levels and bring on cardiac _____.

ANSWER sodium ... potassium ... arrhythmias

QUESTION 347 In both acute and chronic renal failure, imbalanced retention or excretion of both _____ and _____ can occur, bringing on either edema or hyperkalemia.

ANSWER sodium ... potassium

QUESTION 348 Reduced ammonia synthesis in renal failure causes _____ _____ build-up, making the patient _____.

ANSWER hydrogen ion ... acidotic

QUESTION 349 When a kidney patient becomes _____, the body retains water and nitrogenous metabolites, bringing on _____ and _____.

ANSWER oliguric ... edema ... uremia

SUMMARY OF UNIT FIVE

During renal failure, the standard temporary measure for preserving balanced concentrations of extracellular fluids is _____.

QUESTION 350

hemodialysis

ANSWER

In diabetes, cells are unable to utilize _____ for energy because of a lack of _____; to meet energy requirements, the diabetic metabolizes _____ and _____.

QUESTION 351

glucose . . . insulin . . . fats . . . protein

ANSWER

In the diabetic, accumulation of acid _____ bodies in the blood increases hydrogen ion concentration, causes depletion of _____, and generates a fall in _____.

QUESTION 352

ketone . . . sodium . . . pH

ANSWER

Signs of diabetes include (1) significantly higher levels of _____ in the blood; (2) presence of _____ in the urine; (3) polyuria, or _____ urinary volume; and (4) polydipsia, or _____.

QUESTION 353

glucose . . . glucose . . . increased . . . excessive thirst

ANSWER

The standard treatment for diabetes, administration of insulin, is for the purpose of restoring utilization of _____ for _____ needs.

QUESTION 354

glucose . . . energy

ANSWER

Emphysema is a lung condition in which functional respiratory membrane is reduced because of ruptured alveolar _____, enlarged _____ spaces, and destroyed _____ supplying the pulmonary circuit.

QUESTION 355

walls . . . air . . . capillaries

ANSWER

Among the adverse effects of emphysema on fluid, electrolyte, and acid-base balance are (1) _____ failure from back pressure on the right ventricle, (2) respiratory _____ from carbon dioxide and hydrogen ion retention, and (3) systemic _____ from the heart's inability to propel blood.

QUESTION 356

heart . . . acidosis . . . edema

ANSWER

322 CLINICAL CORRELATIONS

QUESTION 357 Therapy for emphysema aims at making the remaining undamaged respiratory apparatus as functional as possible by use of (1) _____ to combat infection and (2) _____ and _____ _____ breathing to improve respiration.

ANSWER antibiotics ... bronchodilators ... positive pressure

QUESTION 358 The most superficial burns are _____-degree burns; the most serious burns, _____-degree.

ANSWER first ... third

QUESTION 359 Third-degree burns are critical if they cover over _____ per cent of the body surface; second-degree, if over _____ per cent of the body surface; first-degree, if over _____ per cent of the body surface.

ANSWER 10 ... 30 ... 50

QUESTION 360 The principal fluid and electrolyte problems associated with burns are (1) _____ formation, (2) _____ loss from denuded areas of the skin, (3) depletion of total body _____, and (4) hyperkalemia due to release of _____ from within cells into extracellular fluid.

ANSWER edema ... protein ... sodium ... potassium

QUESTION 361 In the second phase of recovery from burns, blood volume _____, sodium is _____ with water, and serum levels of potassium are _____.

ANSWER increases ... excreted ... depleted

QUESTION 362 Signs of the first phase of recovery from burns and signs of the second phase of recovery are exactly _____.

ANSWER reversed (opposite)